Operating a Distillation System

By Steve Licht

Distillation Without Equations

First published by AuthorHouse 09/12/05

ISBN: 1-4208-8035-7 (sc)

Printed in the United States of America
Bloomington, Indiana

This book is printed on acid-free paper.

Book Design and Illustrations by Jacqueline Licht

Operating a Distillation System

Part of the
Distillation Without Equations
Technical Seminar

By Steve Licht

Mastering the Art of Distilling

- Taking over watch of an already running distillation system:
    - What's going on inside?
    - What am I supposed to do, or not do?
- Problems that might come along:
    - What to watch for & what to do about it
- How to shut down & how to start up again
- How to keep the distillation operation on course:
    - How to monitor quality & equipment behavior

Suppose that you are assigned to operate this distillation system, already running

Or, suppose this new distillation system arrived at work, some engineers started it going, & then they turned it over to you!

No Problem!

- Taking watch over the operation of an already running distillation system is no problem
- As long as everything keeps going smoothly. . . No Problem!

Let's discuss what is going on inside this distillation system, before we learn to stop or start it, or how to handle problems

- What is turned on?
- What is flowing?
- What is heating?
- What is boiling?
- What is condensing?
- What is cooling?
- What is contacting?
- What is pumping?
- What is coming in & going out?
- What requires control?
- What needs monitoring & recording?
- What needs sampling?
- What other things are standing by, just in case?

What electrical devices, utilities, & equipment are turned on?

- Control panel power
- Pump motor power
- Agitator motor power
- Instrument air
- Cooling source
- Heating source
- Fire protection
- Phone, radio, security

Control panel devices requiring power supply turned on:

- Panel AC & DC
- UPS input & output
- PLC or DCS
- HMI or SCADA
- PC operator stations
- Alarm panels
- Analyzers
- Printers

Motor Control Center (MCC) devices turned on:

- MCC 3-phase supply
- Motor starters for all motors in use
- Breakers to instrument power transformers
- Breakers to lighting panels
- Breakers to UPS feeds

Pumps & agitator motors switched on - possible ways:

- On/off switch at MCC
- Local on/off switch by motor or on panel
- Local hand/off/auto switch by motor or in panel
- Automatic run from control panel relay logic
- Manual run from HMI
- Automatic run from PLC or DCS system

Compressed air supply for instruments turned on:

- Instrument air compressors running
- Instrument air dryer working
- Air supply pressure OK
- Instrument air filters OK
- Air valves on headers & branch lines to instruments are open
- Air bleed & drain valves are closed

What is flowing?

- A hot fluid is providing heat
- A cold fluid is providing cooling
- Feed material is flowing in
- Purified distilled product is flowing out
- Waste streams are flowing out
- Column bottoms are being boiled
- Column overheads are being condensed
- Part of the condensed overhead flows back to the top of the column as liquid reflux
- Gases that cannot be condensed are flowing out

What is flowing inside of the system?

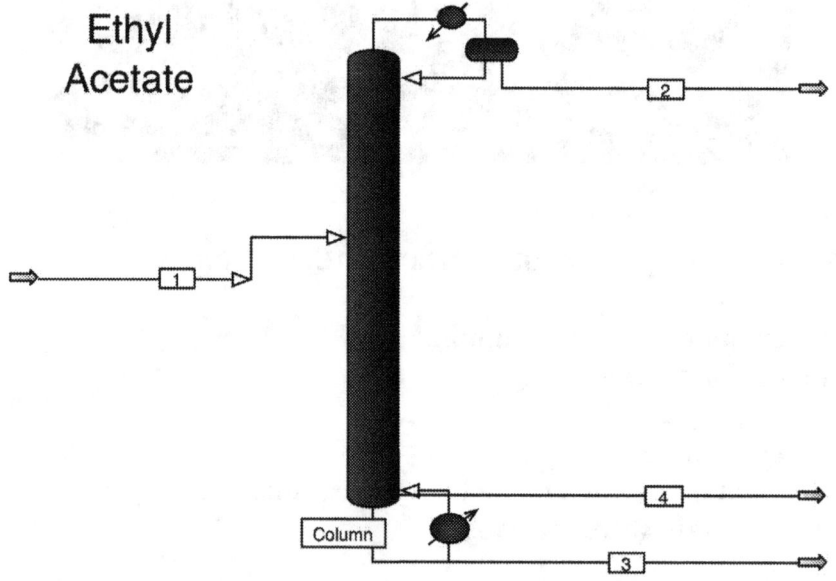

Ethyl Acetate system flow: 1: Feed 4: Product 2,3: Wastes

Stream Name	1	2	4	3
Stream Description	Feed	OVHDS Waste	Main Product	CRUD
Phase	Liquid	Liquid	Vapor	Liquid
KG/HR	2310.000	1751.617	548.383	10.000
Temperature C	10.000	64.721	82.587	82.910
Pressure BAR	4.000	1.000	1.209	1.220
Molecular Wight	69.989	65.710	87.950	88.086
Weight Comp. Percents				
H_2O	4.3290	5.7033	0.0182	0.0007
Ethanol	6.2771	8.2564	0.0689	0.0093
EOAC	72.6407	63.9912	99.7706	99.9513
IPA	0.5195	0.6659	0.0611	0.0146
MEAC	16.2338	21.3832	0.0811	0.0241
Weight Comp. Rates KG/HR				
H_2O	100.0000	99.9000	0.1000	0.0001
Ethanol	145.0000	144.6210	0.3780	0.0009
EOAC	1677.9999	1120.8801	547.1246	9.9951
IPA	12.0000	11.6632	0.3353	0.0015
MEAC	374.9999	374.5526	0.4450	0.0024
Enthalpy M*KCAL/HR	0.031	0.077	0.069	0.000

Ethyl Acetate column internal flows - Graphical representation

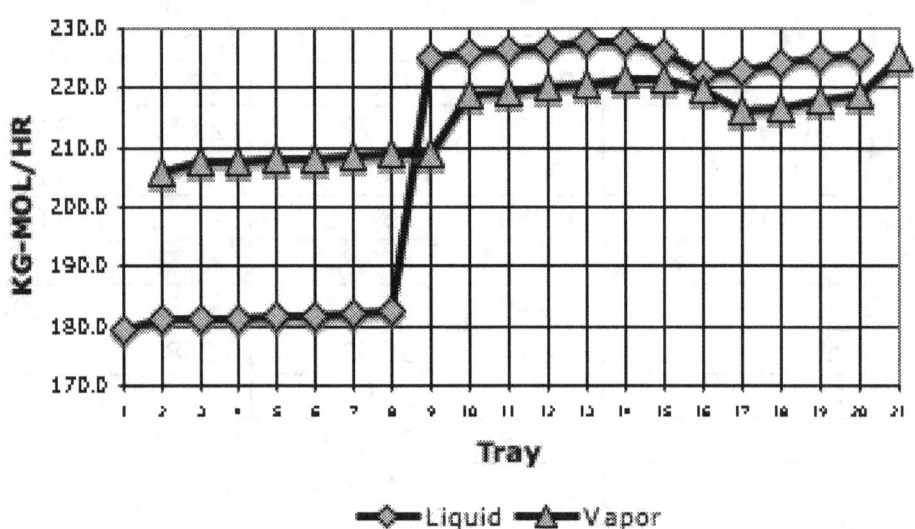

What is heating?

Heat is flowing energy

- Heat is the flow of energy from hot to cold material
- Objects don't contain heat
- Heat flow = Heat transfer
- "Specific Heat" is the same as "Heat Capacity":
  - Both mean how much heat is needed to raise the temperature of one pound of something by one degree

Heat transfer

- Just means that heat goes from one material (hot) to another material (cold)
- Heat can be transferred three ways:
  1. Conduction
  2. Convection
  3. Radiation

Heat transfer by conduction:

- Is the way heat flows through solids
- Is the way heat flows through motionless liquids
- Is very, very slow through gases & through air pockets inside of bulk solids or bulk liquids
- Heat always flows from higher to lower temperature places
  - [400°F -->Heat flows this way--> 396°F]
- Conductors (metals) are good at conducting heat
- Insulators (insulation) conduct heat poorly

Heat transfer by convection:

- Means that warmed or cooled fluid moves from one place to another
- **Natural Convection** means the fluid is not being stirred or pumped
  - Teapot on stove, tank with cooling coil
- **Forced Convection** means the fluid moves by stirring or pumping
  - A stirred pan, jacketed reactor, shell & tube heat exchangers

Heat transfer by radiation:

- Is what sunshine is
- Is heat radiating from a fire
- Is what is meant by "red-hot"
- Is what goes on in the fireboxes in boilers, furnaces, & fired heaters
- Radiation can go through empty space
- Radiated heat can either be reflected (mirror), transmitted (clear glass), or absorbed

Ethyl Acetate column top to bottom temperature graph

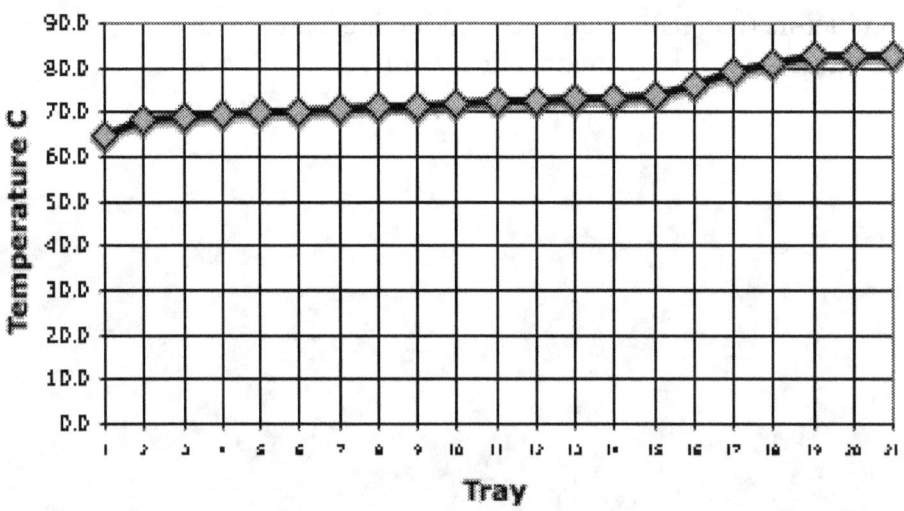

What is boiling?

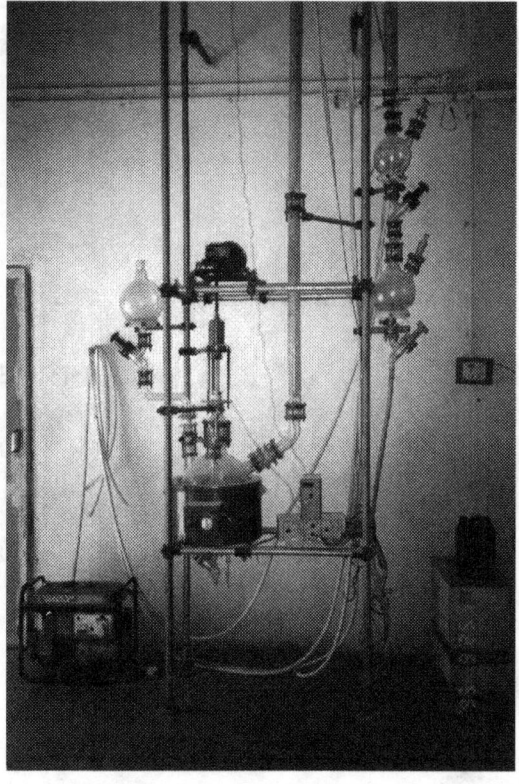

Boiling

- Means adding heat to a liquid & making it vaporize
- The liquid must be in contact with a hot surface which is hotter than the boiling point of the liquid at whatever pressure there is in the "boiler" or "reboiler"
- In distillation systems, the same material goes back to be boiled again & again, hence the traditional term "reboiler"

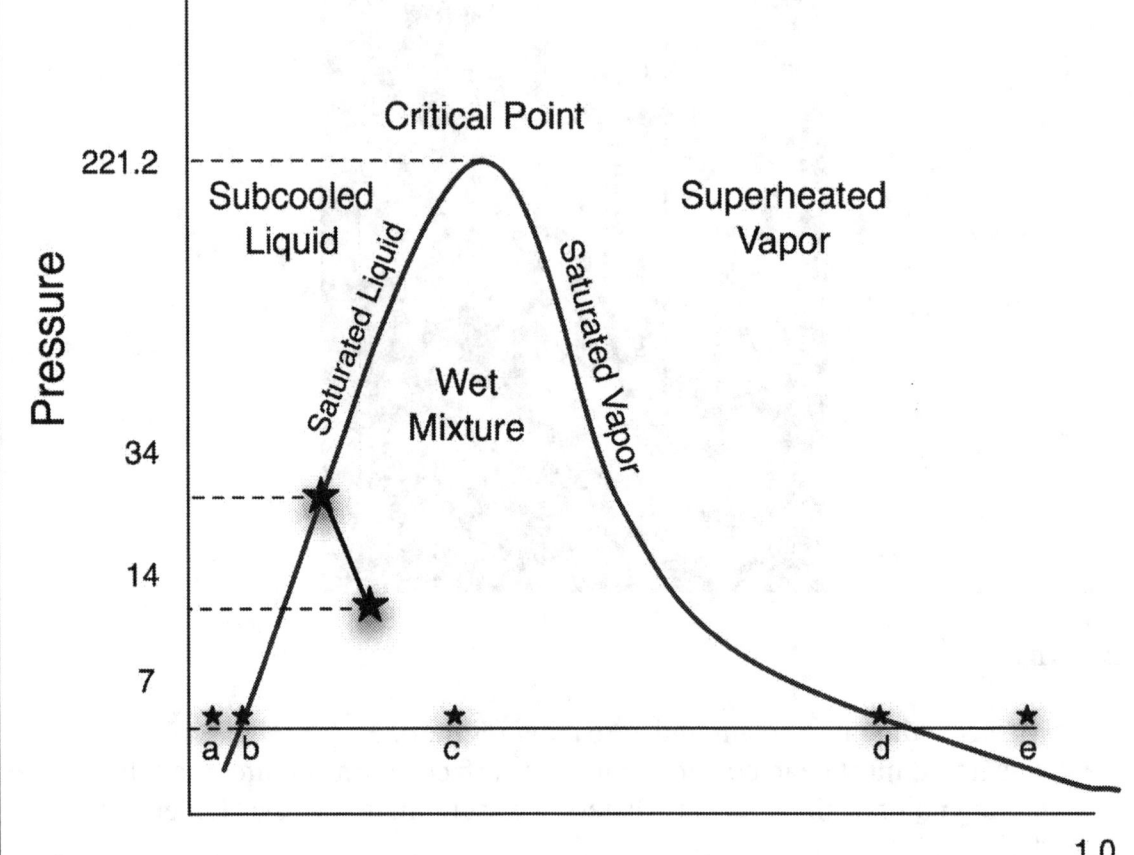

Plot of Pressure vs Volume for a Liquid

Critical Point

221.2

Subcooled
Liquid

Saturated Liquid

Superheated
Vapor

Saturated Vapor

Wet
Mixture

Pressure

34

14

7

a b   c   d   e

1.0

Specific Volume

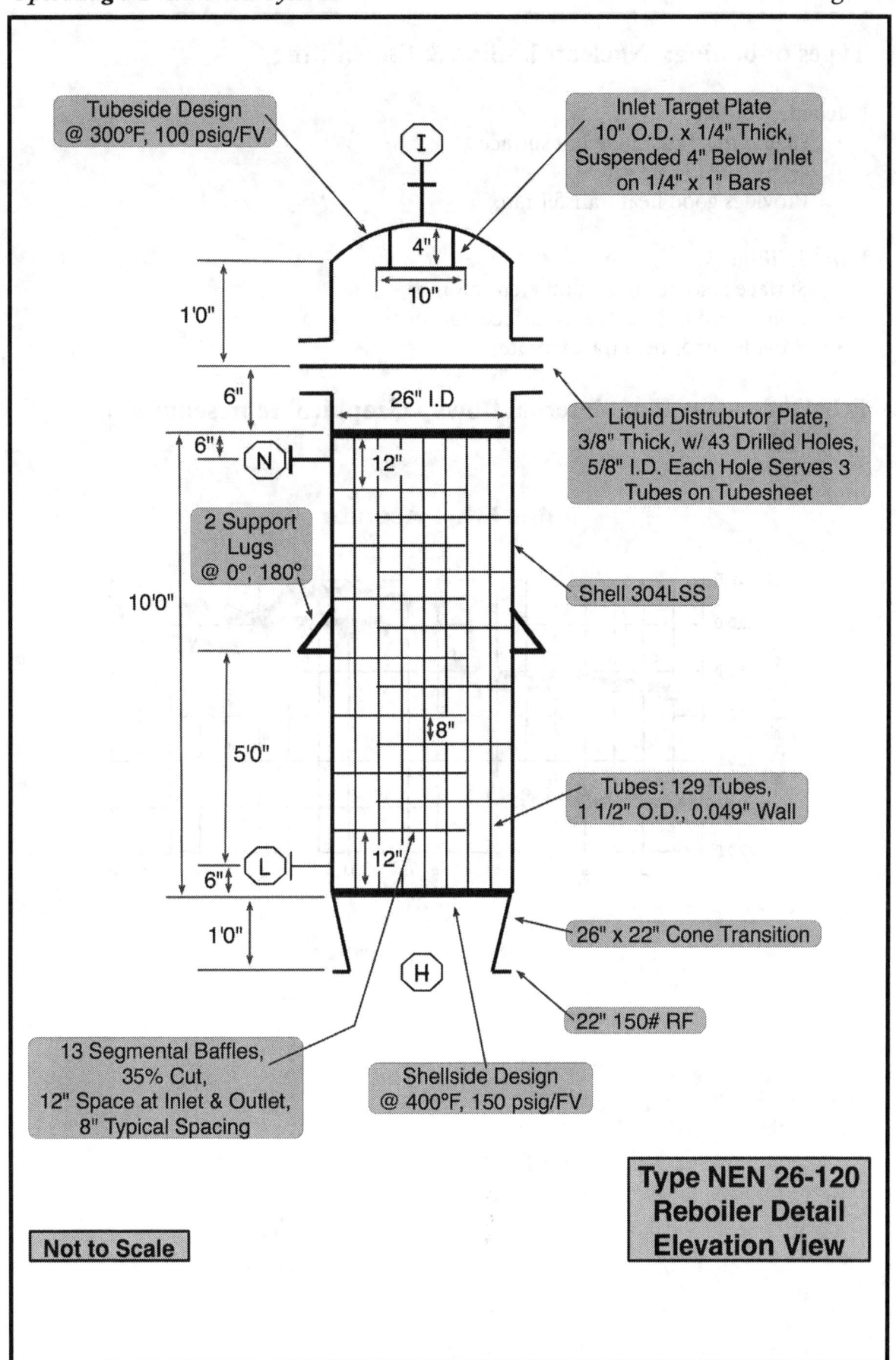

Tubeside Design @ 300°F, 100 psig/FV

Inlet Target Plate 10" O.D. x 1/4" Thick, Suspended 4" Below Inlet on 1/4" x 1" Bars

4"

10"

1'0"

6"

26" I.D

6"

N

12"

Liquid Distrubutor Plate, 3/8" Thick, w/ 43 Drilled Holes, 5/8" I.D. Each Hole Serves 3 Tubes on Tubesheet

2 Support Lugs @ 0°, 180°

Shell 304LSS

10'0"

8"

5'0"

Tubes: 129 Tubes, 1 1/2" O.D., 0.049" Wall

12"

6"

L

1'0"

26" x 22" Cone Transition

H

22" 150# RF

13 Segmental Baffles, 35% Cut, 12" Space at Inlet & Outlet, 8" Typical Spacing

Shellside Design @ 400°F, 150 psig/FV

Type NEN 26-120 Reboiler Detail Elevation View

Not to Scale

Types of boiling: Nucleate boiling & film boiling

Nucleate Boiling:
- Liquid touches all the hot surface
- Considered more safe
- Provides good heat transfer rate

Film Boiling:
- Surface is so hot it is smothered in vapor
- Considered unsafe due to surface hot spots
- Provides poor heat transfer rate

Ethyl Acetate column internal flows - Graphical representation

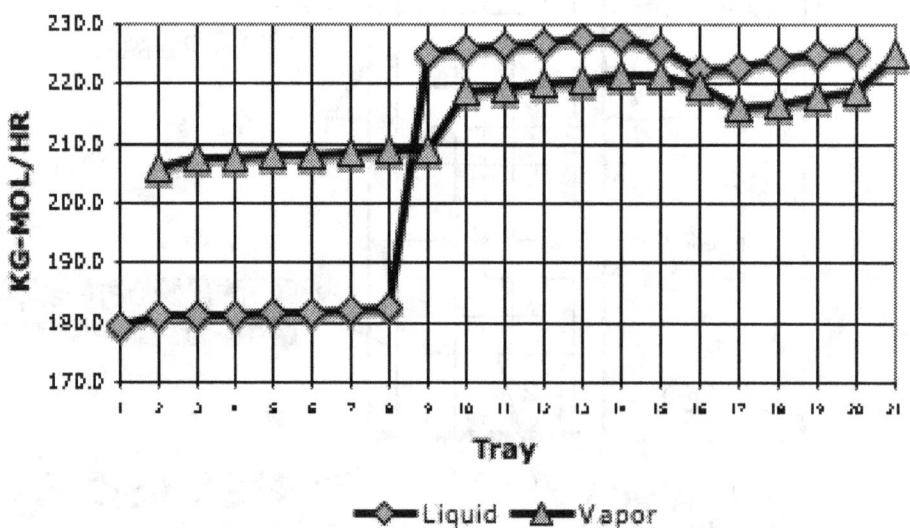

What is condensing?

Condensation is the opposite of boiling

- Condensation means removing heat from a vapor & making it turn back into a liquid
- The vapor must be in contact with a cold surface which is colder than the boiling point (also called the dew point) of the vapor at whatever pressure there is in the "condenser"
- Condensers usually are shell & tube heat exchangers

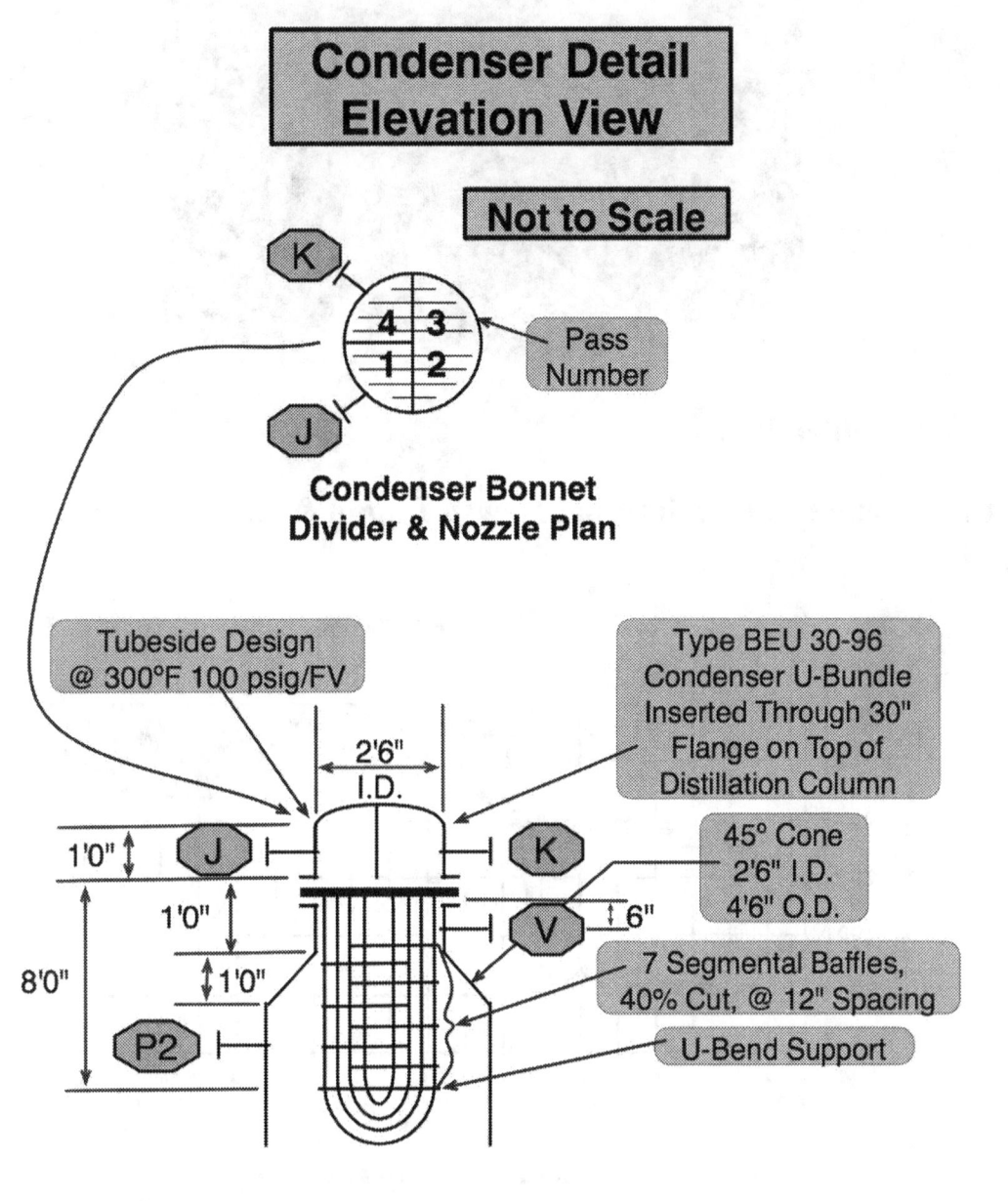

Condenser Detail Elevation View

Not to Scale

Pass Number

Condenser Bonnet Divider & Nozzle Plan

Tubeside Design @ 300°F 100 psig/FV

Type BEU 30-96 Condenser U-Bundle Inserted Through 30" Flange on Top of Distillation Column

2'6" I.D.

45° Cone 2'6" I.D. 4'6" O.D.

6"

7 Segmental Baffles, 40% Cut, @ 12" Spacing

U-Bend Support

1'0"

1'0"

1'0"

8'0"

What is cooling?

What is contacting?

Ethyl Acetate column internal flows: Liquid & vapor by stage

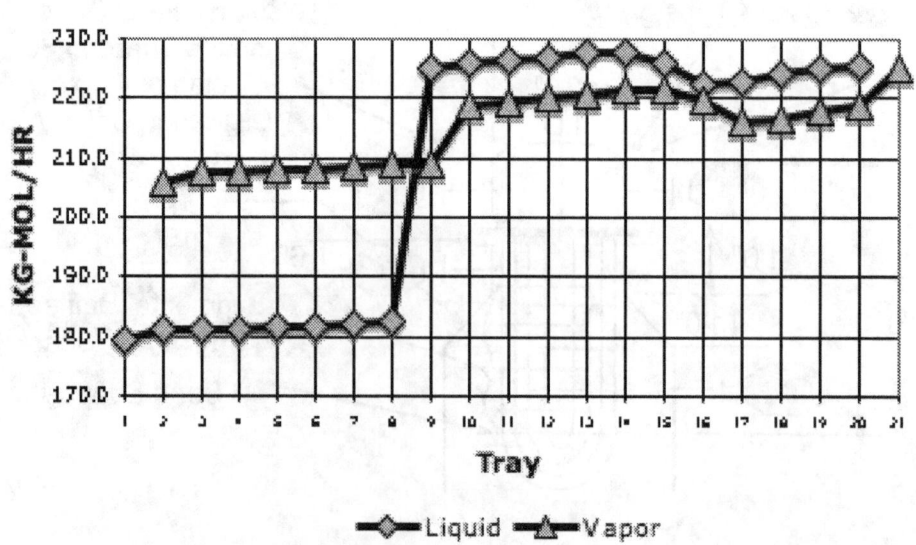

The purity changes stage-by-stage as vapor contacts liquid

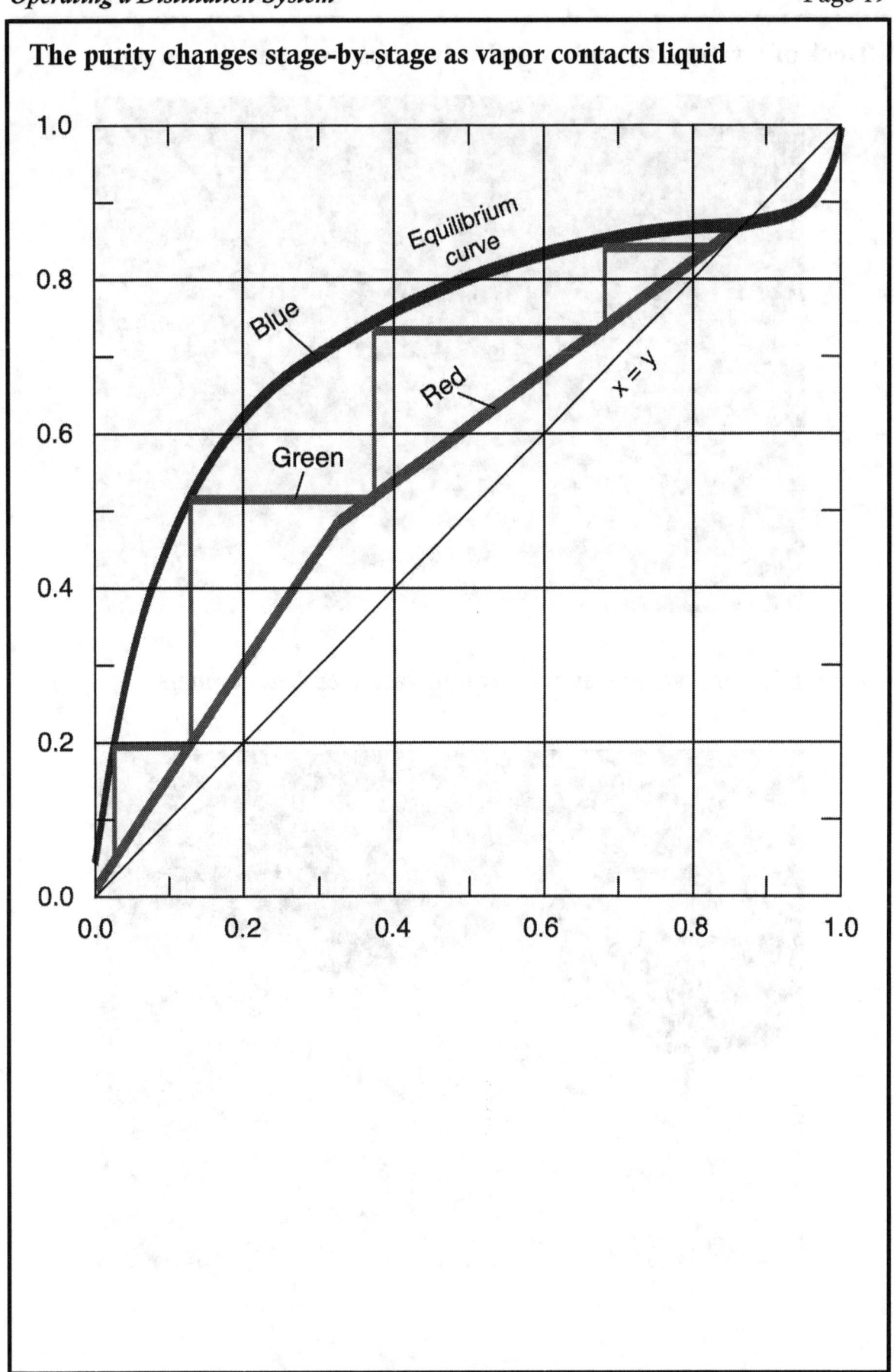

Deck of a valve tray with round valve caps, two pass liquid path

Underside of a valve tray with round valves & downcomers

Trays are mounted horizontally inside distillation columns

One part of multi-section valve tray with rectangular valves

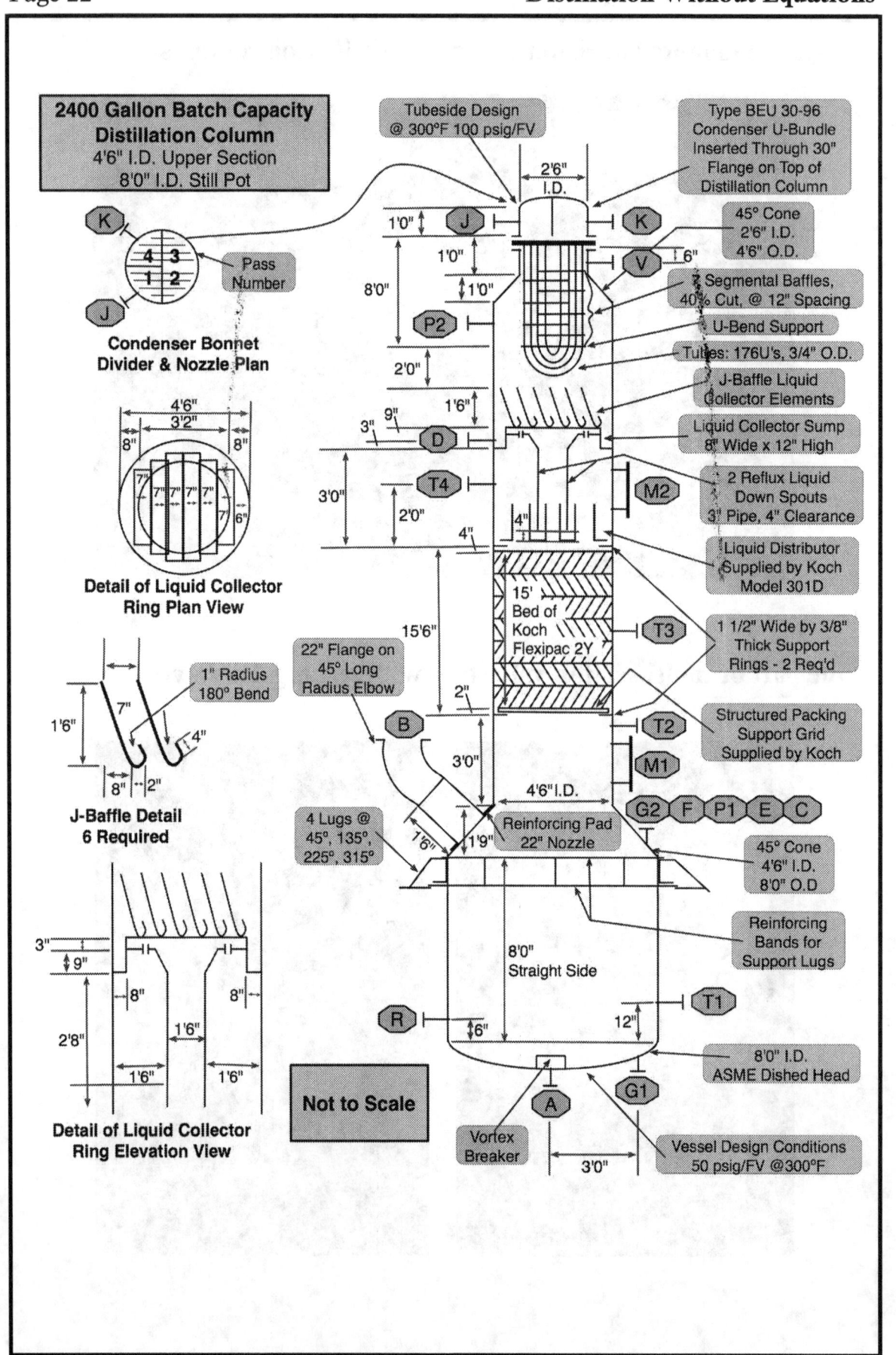

2400 Gallon Batch Capacity Distillation Column
4'6" I.D. Upper Section
8'0" I.D. Still Pot

Condenser Bonnet
Divider & Nozzle Plan

Pass Number

4 3
1 2

Detail of Liquid Collector
Ring Plan View

J-Baffle Detail
6 Required

1" Radius
180° Bend

Detail of Liquid Collector
Ring Elevation View

Not to Scale

Tubeside Design
@ 300°F 100 psig/FV

Type BEU 30-96
Condenser U-Bundle
Inserted Through 30"
Flange on Top of
Distillation Column

45° Cone
2'6" I.D.
4'6" O.D.

Segmental Baffles,
40% Cut, @ 12" Spacing

U-Bend Support

Tubes: 176U's, 3/4" O.D.

J-Baffle Liquid
Collector Elements

Liquid Collector Sump
8" Wide x 12" High

2 Reflux Liquid
Down Spouts
3" Pipe, 4" Clearance

Liquid Distributor
Supplied by Koch
Model 301D

1 1/2" Wide by 3/8"
Thick Support
Rings - 2 Req'd

15'
Bed of
Koch
Flexipac 2Y

Structured Packing
Support Grid
Supplied by Koch

22" Flange on
45° Long
Radius Elbow

4 Lugs @
45°, 135°,
225°, 315°

Reinforcing Pad
22" Nozzle

45° Cone
4'6" I.D.
8'0" O.D.

Reinforcing
Bands for
Support Lugs

8'0"
Straight Side

8'0" I.D.
ASME Dished Head

Vortex
Breaker

Vessel Design Conditions
50 psig/FV @300°F

What is pumping?

There are different types of pumps for different duties

Positive displacement pumps:

- Cannot have flow restricted, while they are running - NEVER
- Can maintain a fixed flow rate, despite variations in back pressure
- Can be much easier to prime than centrifugal pumps
- Useful for transfer pumps that must be frequently drained

Centrifugal pumps:

- Can vary in flow rate across a wide range
- Any change in back pressure changes the flow rate
- Can be cut back to zero flow for approximately one minute without damage
- One hour at zero flow will overheat & damage a centrifugal pump

Figure 23
Cast iron pump casing
fractured into sections
due to pressure pulsa-
tions.

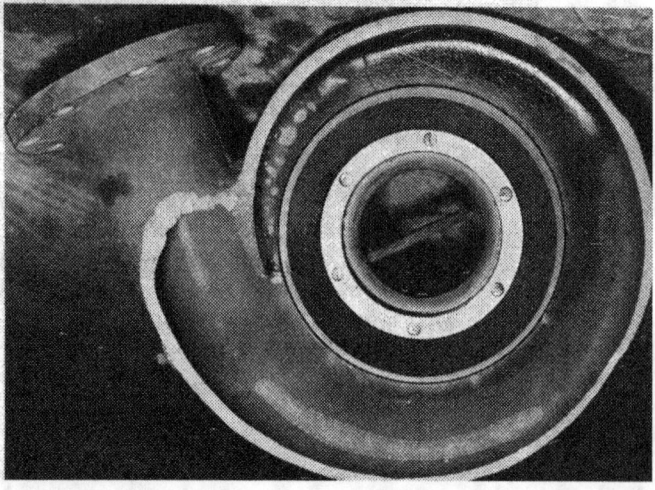

What is coming in?

IPA distillation 1: Feed 2: Waste 3: Product

IPA/Water			
Stream ID	1	2	3
Temperature (C)	15.0	81.9	82.8
Mass Flow (KG/HR)	800.000	150.924	649.079
Mass Fraction			
Water	0.002	0.010	100 PPM
ISOPR-01	0.998	0.990	1.000
METHA-01			
DIBUT-01			

What is going out?

IPA / Methanol distillation system 1: Feed 2,4: Waste 3: Product

IPA/Methanol/Water				
Stream ID	1	2	3	4
Temperature (C)	15.0	66.3	82.8	74.6
Mass Flow (KG/HR)	500.000	27.000	446.000	27.000
Mass Fraction				
Water	0.002	0.008	102 PPM	0.028
ISOPR-01	0.948	0.249	1.000	0.790
METHA-01	0.050	0.744	918 PPB	0.182
DIBUT-01				

What requires control?

A control strategy (manual or automatic) is required for the following:

- Feed input
- Heat input
- Column product purity control
- Column bottom effluent
- Reflux
- Column top distillate
- Other feed, product, or internal streams
- System pressure

Manually or automatically set control valves carry out the selected control strategy

What needs monitoring & recording?

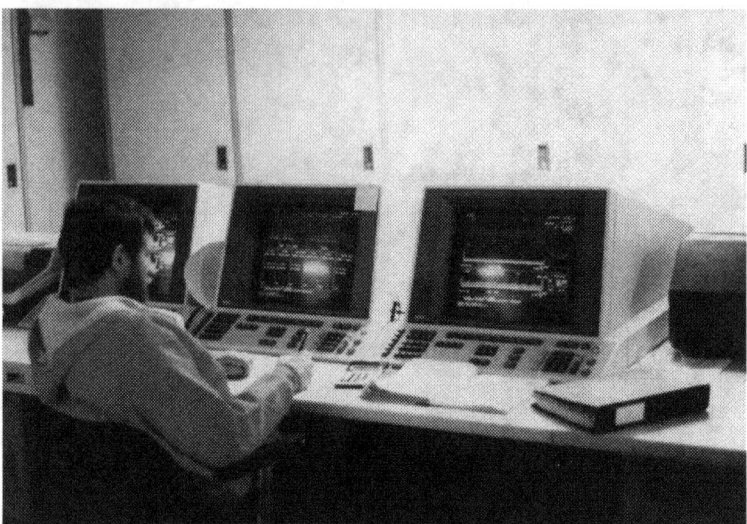

In addition to control loops, process indicators & safety interlocks require monitoring & recording

- Indicators are for operator information or for data recording purposes
- Interlock devices monitor their inputs & respond if required for protection of pumps & overall system safety

What needs sampling?

IPA/Methanol/Water				
Stream ID	1	2	3	4
Temperature (C)	15.0	66.3	82.8	74.6
Mass Flow (KG/HR)	500.000	27.000	446.000	27.000
Mass Fraction				
Water	0.002	0.008	102 PPM	0.028
ISOPR-01	0.948	0.249	1.000	0.790
METHA-01	0.050	0.744	918 PPB	0.182
DIBUT-01				

What other things are standing by, just in case they are needed?

Safety equipment is standing by:

- Pressure safety valve (PSV) or burst disk pressure safety element (PSE) relieves pressure in each section of the system containing a vessel that can be blocked in
- High level switches & low level switches
- High pressure switches
- Interlocks triggered by switches or alarms

Problems that might come along

What to watch for . . .

& what to do about it!

Problems to watch for

Problems that might occur while a distillation system is operating can be sorted into categories:
- Quantity problems
- Quality problems
- Integrity problems
- Safety problems

Quantity problems can refer to quantity of feed, product, wastes, utilities, containers, lab services

- The quantity of available feed or the feed flow rate may increase or decrease
- There may be too much or not enough product or waste materials coming out of the system
- There may be too little or excessive flow of steam, hot oil, cooling water, glycol, nitrogen, air, etc.
- There may not be enough tanks or drums to hold products or wastes, or vials to hold samples
- The lab might not be able to analyze samples as fast as demanded

Figure 44
The sphere at the left (which contained butane) was
pumped over. The spilled butane ignited and the
resulting fire ruptured the sphere shell.

What to do about quantity problems

- Most distillation systems continue operating fine with any feed rate within a broad range
- Not enough or too much product or waste flow should always be investigated until understood
- Insufficient utilities eventually lead to quality problems or safety problems if not restored
- There are always more product containers somewhere - slow down production as last resort
- If the lab is overloaded, negotiate priorities or help them by doing analyses yourself if you can

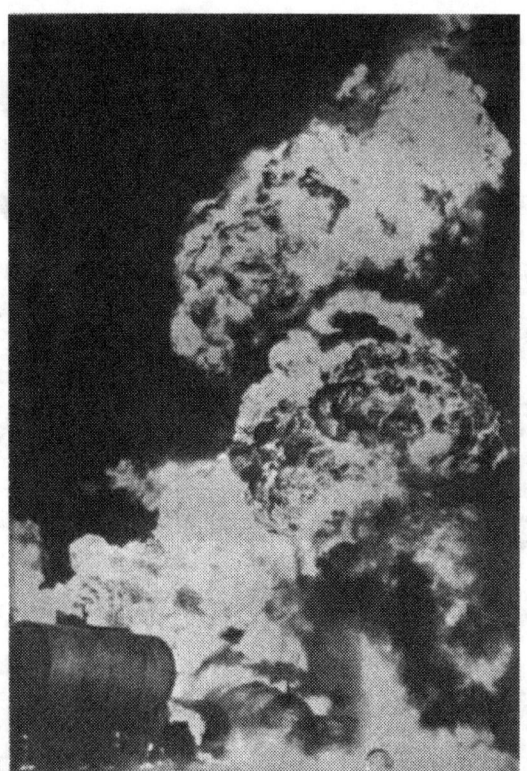

Figure 5   Fargo, N.D.
This loading rack fire resulted when
the truck driver forgot to disconnect
his fill hose before driving away.

Quality problems can refer to feed, product, wastes, utilities, lab services

- Feed quality too high will require changes in the operation, possibly lower feed rate
- Feed quality too low will require changes in operation, possibly higher feed rate
- Waste product quality deviations require attention to be certain product is not lost
- Utility quality deviations require monitoring, because they may lead to other problems
- Lab service quality problems may require technical or organizational solutions

Integrity problems are more serious & will not go away by themselves if ignored

- No visible leaks should be tolerated other than minor quantities of steam, water, air
- Flange gaskets, threaded connections, & valve stem packing do require occasional tightening
- Broken supports should always be fixed in an improved manner not to break again
- Noisy or vibrating equipment or piping should have maintenance attention soon
- Chemical leaks or hazardous utility leaks require quick action to stop the leak & repair the source

Figure 20

An electrical short was the source of ignition which destroyed this barrel house.

Safety problems can be little or big, but little problems left uncorrected tend to get bigger

- Steam leaks get bigger & become burn hazards or burst hazards
- Chemical leaks at flanges & valve packing become toxic or explosive vapor clouds or slick corrosive or flammable liquid pools
- Broken supports lead to more broken supports & eventually something falling
- Noisy bearings, pump seals & couplings, & electric motors never heal themselves

Figure 3

Pump stuffing box leakage, probable electrical insulation failure and air in this pump room completed the fire triangle with explosive force. Adequate ventilation and well-maintained equipment will eliminate such accidents.

Major safety problems demand action as they threaten to become true emergencies

- Any loud or repeating snapping or cracking sound demands emergency shutdown & all personnel cleared from the area
  - Alert supervisors & do not enter the suspicious area alone
- Any large & quick pressure spike demands emergency shutdown & venting.
- Steadily rising temperature measurements require treatment as if a fire already is burning.
- Flames from solvent fires often are not visible in daylight
  - Never walk into any area where solvent is spilled or a solvent fire is suspected

Figure 21

Breakage of a glass enclosed "look box" released light naphtha. Vapors spread to a hot tar line where they ignited and flashed back, exploding other combustible mixtures in the area.

Emergency handling: Intervention & shutdown

- If safety interlocks or emergency systems are triggered, do not prevent them from acting
- When there is an emergency in a system equipped with emergency stop (E-Stop) buttons, use them
- If required to intervene manually, first block any energy inputs to the system (steam, hot oil), then shut off all electric motors, then block all material inputs to the system
- It is usually better to let pressure safety valves & burst disks relieve system pressure, than to open vent valves to atmosphere during an emergency

Figure 4
Fire at refinery storage area

How to make a smooth shutdown

- Ramp the feed flow rate down to zero
- Ramp the heat input flow rate down to zero
- Keep all other instruments functioning, control loops in automatic, pumps running, & utilities flowing
- As material output flows decrease naturally, close their flow paths & stop pumps
- When all pumps have stopped, & vessels contain only safe residual liquid heels, vent or inert the system as prescribed for shutdown periods

When & why to empty a distillation system

- A distillation system should be emptied when the next production material is not compatible with the liquid heel left inside after a shutdown
- Some or all of the equipment or piping may need to be emptied & blinded off to make it safe for work to be done during the shutdown period
- Drain all pipelines to vessels or to drain connections, pump all vessels empty, blow out lines with nitrogen or steam, water rinse, air dry
- Do not work inside nitrogen blanketed equipment

How to fill an empty distillation system to prepare for start up

- First make certain that no one is still working on or inside of equipment
- Make certain that all equipment & piping is in place, blinds are removed, drains are closed, & normally open manual valves are open
- Only the distillation column base & reboiler circulation loop need to be filled to normal liquid levels before a start up
- Other equipment such as the decanter & reflux drum can remain empty at this time

How to start up a cold (but not empty) distillation system

- Make certain that all cooling systems & vent collection systems are working
- With a forced circulation reboiler, start the reboiler circulation pump
- Slowly ramp open the heat source to the reboiler (steam or hot oil) to near normal heat input
- When material begins boiling & the column level decreases, add more feed to top up the column bottom as required, whenever required
- When the column is hot all the way to the top, condensate will appear in the reflux equipment

How to line out a newly started or restarted distillation system to achieve a total reflux condition

- When condensate appears in the reflux equipment, start the reflux pump, allowing all of the reflux to return to the column
- Keep topping up with feed material as needed
- Ramp the heat input rate up to normal
- When column temperatures line out & no more material top ups are required, the system is ready to begin accepting continuous feed

How to start feed, recycle, & line out a distillation system to normal compositions & flows

- Ramp up feed rate quickly to 75% of normal
- Make certain that control loops for bottoms flow out & distillate out are functioning
- Frequently bottoms & distillate are recycled back to the feed tank until on-spec
- When column temperatures stabilize, increase feed to 100% of normal, & allow to re-stabilize

When & how to recognize that a distillation system has begun to make good product

- All the flow rates in the system will be near their usual values
- All the column temperatures will be near their usual values
- A trend recording of the temperatures & flows will show straight lines
- On-line analyzers & density meters will show values within good quality specifications

How to continue long-term operation with on-spec quality

- Primarily, it was the designers' duty to make certain that the distillation system includes robust control configurations which ensure quality
- If you must control something manually to maintain on-spec quality, increasing the amount of reflux usually improves the distillate quality & increasing the heat input usually improves the bottoms stream quality
- Decreasing the feed rate improves quality at the top & bottom of the column, in most systems
- Flow blockages often show up as quality crashes

What to do when quality measurements fail

- First, resample & reanalyze, to be sure that the quality failed, not the measurement
- If uncertain about what to do to improve quality, decrease the feed rate by 25% as a first step
- If quality improves, gradually bring the feed rate back up towards the normal value
- If quality did not improve, increase the reflux next
- If quality still did not improve, increase the heat input, watching out for the signs of column flooding (high pressure drop across column)
- If nothing you do works, call someone to help out

Counterintuitive actions can achieve or recover quality

- Sometimes the right thing to do, looks at first like you are doing the opposite of the correct thing!
- Decreasing the heat input may improve both top & bottom product quality, if the column is starting to foam or jet-flood
- When the overhead product coming out of the decanter in an azeotropic partially miscible distillation system contains too much water, often the correct action is to add lots of fresh water to the decanter, to sweep away co-solvency compounds & permit better partitioning

There's on-spec product in those distillate receivers. No problem!

What to do about utility variations:

- Quality
- Quantity
- Supply Pressure
- Temperature
- Unsteadiness
- Cycling

What to do about:

- Noises
- Leaks
- Odors
- Vibration
- Degrading Condition of Equipment
- Degrading Condition of Work Area

When to run away . . . and:

- Hit the emergency stop button, hit the fire alarm station, call emergency services, yell loudly!

Remember, distillation columns should only look like rockets, not fly like rockets!

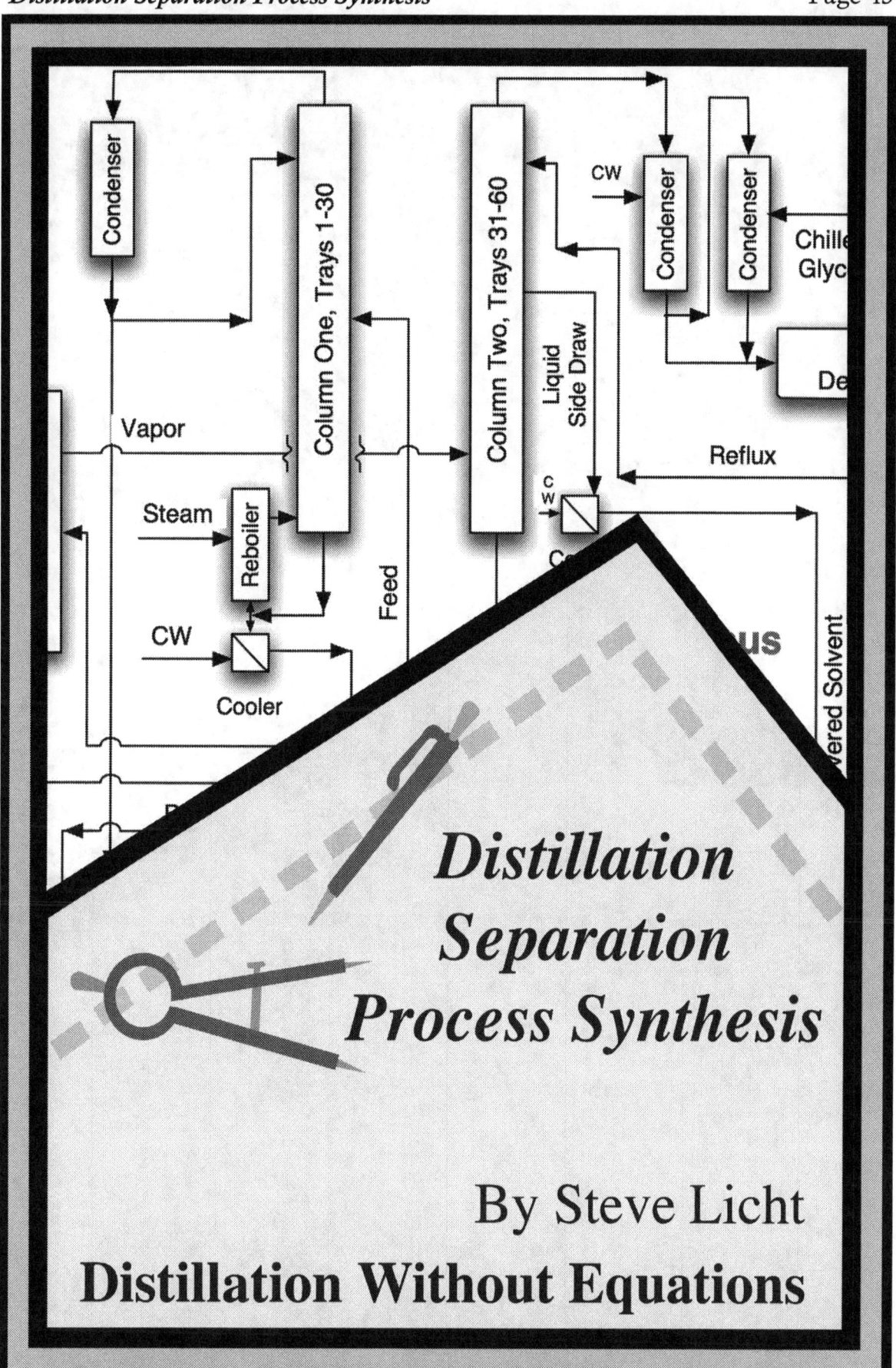

Distillation Separation Process Synthesis

By Steve Licht

Distillation Without Equations

Distillation Separation Process Synthesis

Part of the

Distillation Without Equations
Technical Seminar

By Steve Licht

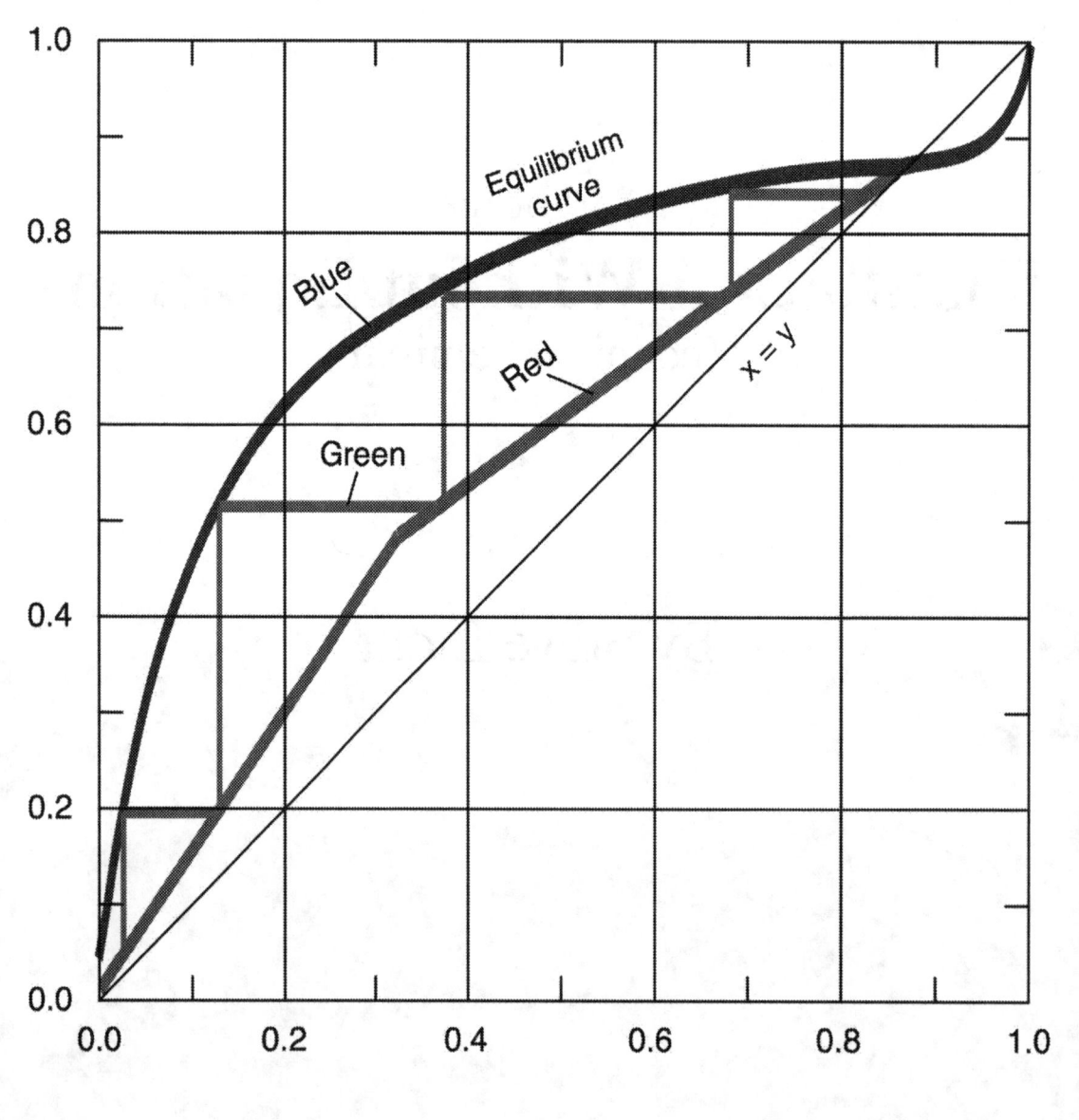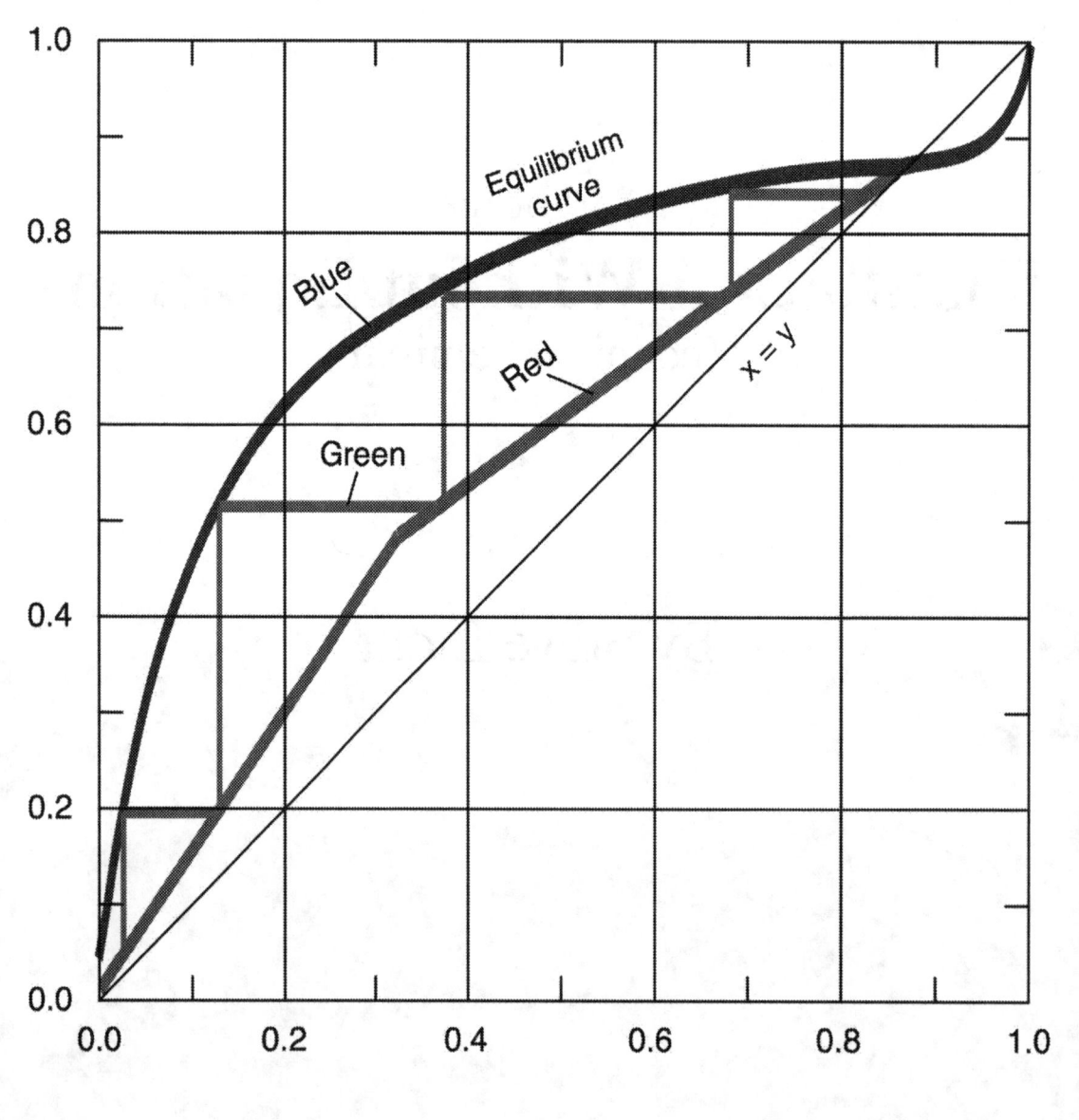

Distillation Separation Process Synthesis

- Actual Vapor Liquid Equilibrium (VLE) relationships in multi-component mixtures
- Advantages & limitations of VLE models
- Equilibrium stage separation process synthesis:
  - How to "un-mix" & purify to required quality
- Flow sheet modeling using computer simulators
- Process optimization:
  - Yield, energy usage, equipment size, simplicity v. complexity
- Laboratory distillation testing to confirm models
- Performance specifications & plant tests

Actual VLE relationships in multi-component mixtures

- Actual multi-component mixture distillation behavior is determined by examining each binary pair of components

This diagram depicts a methylene chloride - methanol binary distillation

- Blue curve represents equilibrium concentration, y=vapor, x=liquid
- The red lines have slopes equal to liquid/vapor molar flow ratio
- Green steps show the change in concentration per distillation stage

The equilibrium curve crosses the x=y line at the azeotrope value, 87% molar methylene chloride

- It is not possible to distill pure methanol from a mixture more than 87% methylene chloride
- It is not possible to distill pure methylene chloride from a mixture with more than 13% molar methanol

The higher the red lines are above the x=y line, reflux is lower, less energy is used, but more stages are required

- Equilibrium curves closer to the x=y axis require more reflux & more separation stages than broadly spaced curves
- The minimum reflux amount occurs where the red line rises up to almost touch the blue line
- Below minimum reflux, your column disappears
  - If the reflux ratio is less than the minimum reflux ratio for this system, the separation of a whole column equals only that of a one stage process

Azeotropes

- Are barriers that can be crossed by adding a third component which raises or lowers the volatility of component #1 or #2

Extractive distillation

- Uses a less volatile third component to carry component #1 down the column, leaving pure #2 in the distillate

Heterogeneous azeotropic distillation

- Uses a more volatile third component (entrainer) to carry component #1 up the column, leaving pure #2 as the column bottoms product

Liquid-liquid extraction

- Can be used to reduce the amount of one component, moving the feed concentration to the other side of the azeotropic point

Combinations of all binary pair VLE relationships

- Fully define the distillation behavior of all non-reacting, non-ionic multi-component mixtures
  - No matter how many components are present!

In a 4 component system, there are 6 binary pairs, as follows:

- Component #1 with #2
- Component #1 with #3
- Component #1 with #4
- Component #2 with #3
- Component #2 with #4
- Component #3 with #4

Binary pair interactions do not fully define VLE in reactive or ionic systems . . . for example:

- Ammonia - water solutions
- Hydrochloric acid solutions
- Amines with water & acid gases
- Acetic acid mixed with amides
- Compounds which form dimers or polymers

In the absence of azeotropes, there is no theoretical limit to the purity that can be achieved by distillation with a sufficiently large number of stages

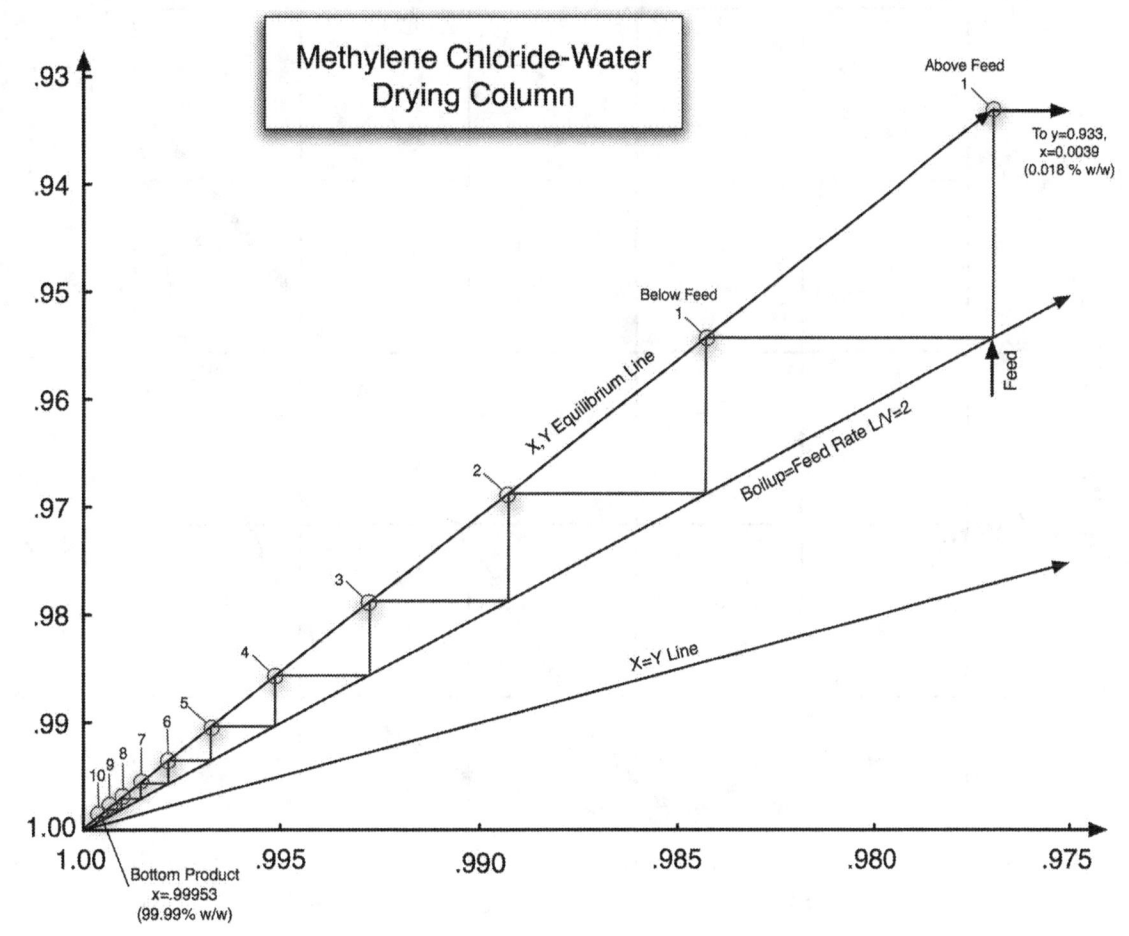

Methylene Chloride-Water
Drying Column

Categories of actual binary pair VLE behavior

- Both pure components may be liquid at the temperature & pressure of the system, or one pure component may be a fixed gas at these conditions
- Component pairs may be azeotropic or non-azeotropic
- Azeotropic pairs may form either a high or a low boiling azeotrope
- The two components may be fully miscible, or when mixed, they may form partially miscible liquid phases or practically immiscible liquid phases
- Either one or both components may exhibit ionic or reactive behavior

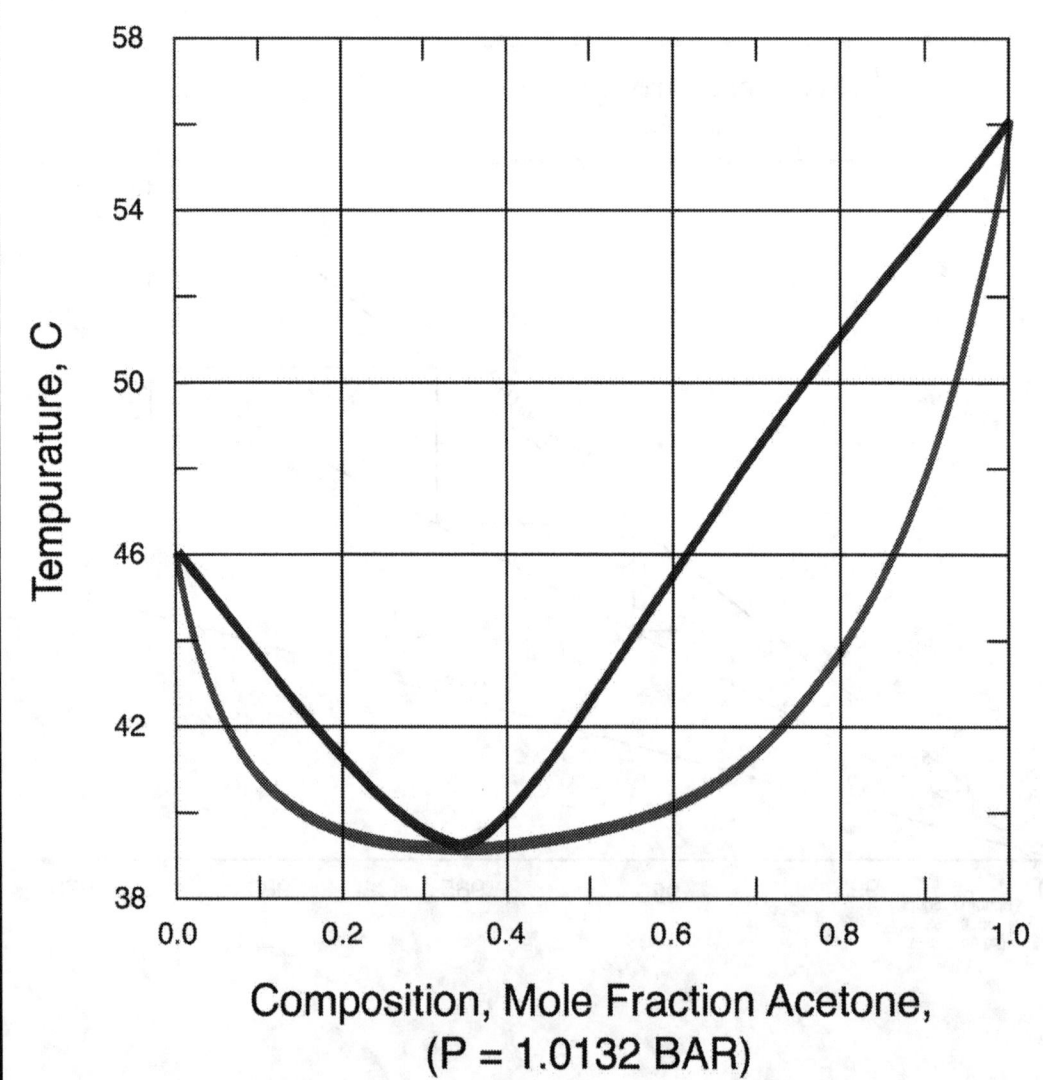

Temperature (T) - liquid mole fraction (X) - vapor mole fraction (Y) diagrams

- TXY & XY diagrams graphically well illustrate VLE behavior of most binary systems
- They are not well suited to systems in which one component is a fixed gas with limited solubility
- These diagrams are not visually informative in practically immiscible liquid systems
- They do not provide sufficient information for determining the behavior of ionic, associative, or reactive components

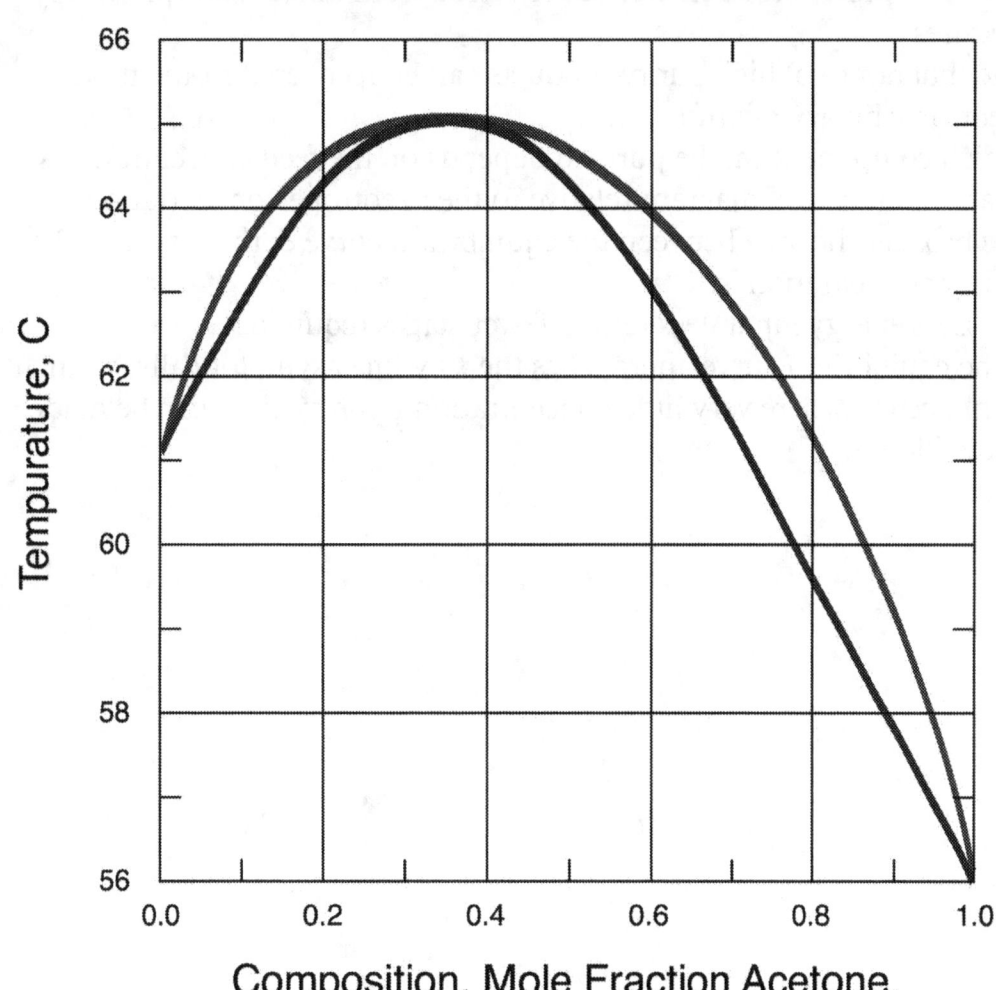

T-X-Y Plot for Acetone and CLFR

XY diagrams determine how to un-mix any two components of a liquid mixture

- The equilibrium curve will tell you, at any point along it, whether a component will tend to go UP the column (y>x) or DOWN the column (y<x)
- If the equilibrium curve crosses the x=y line at any point, there is an azeotrope at the cross-over concentration
- Simple distillation with one feed & two components cannot cross azeotrope barriers

XY diagrams provide a road map to separation scheme synthesis

- Two high purity products can be recovered from non azeotropic binary mixtures
- One (but not two) high purity products can be recovered from an azeotropic binary mixture
- Which component can be purified depends on the feed concentration's location on the XY diagram relative to the azeotrope concentration
- The broader the area between the equilibrium curve & the x=y line, the easier the separation is:
  - Less energy input, less reflux, fewer stages required
- If the equilibrium curve approaches the x=y line asymptotically, a pinch point occurs, where very little concentration improvement can be made by distillation

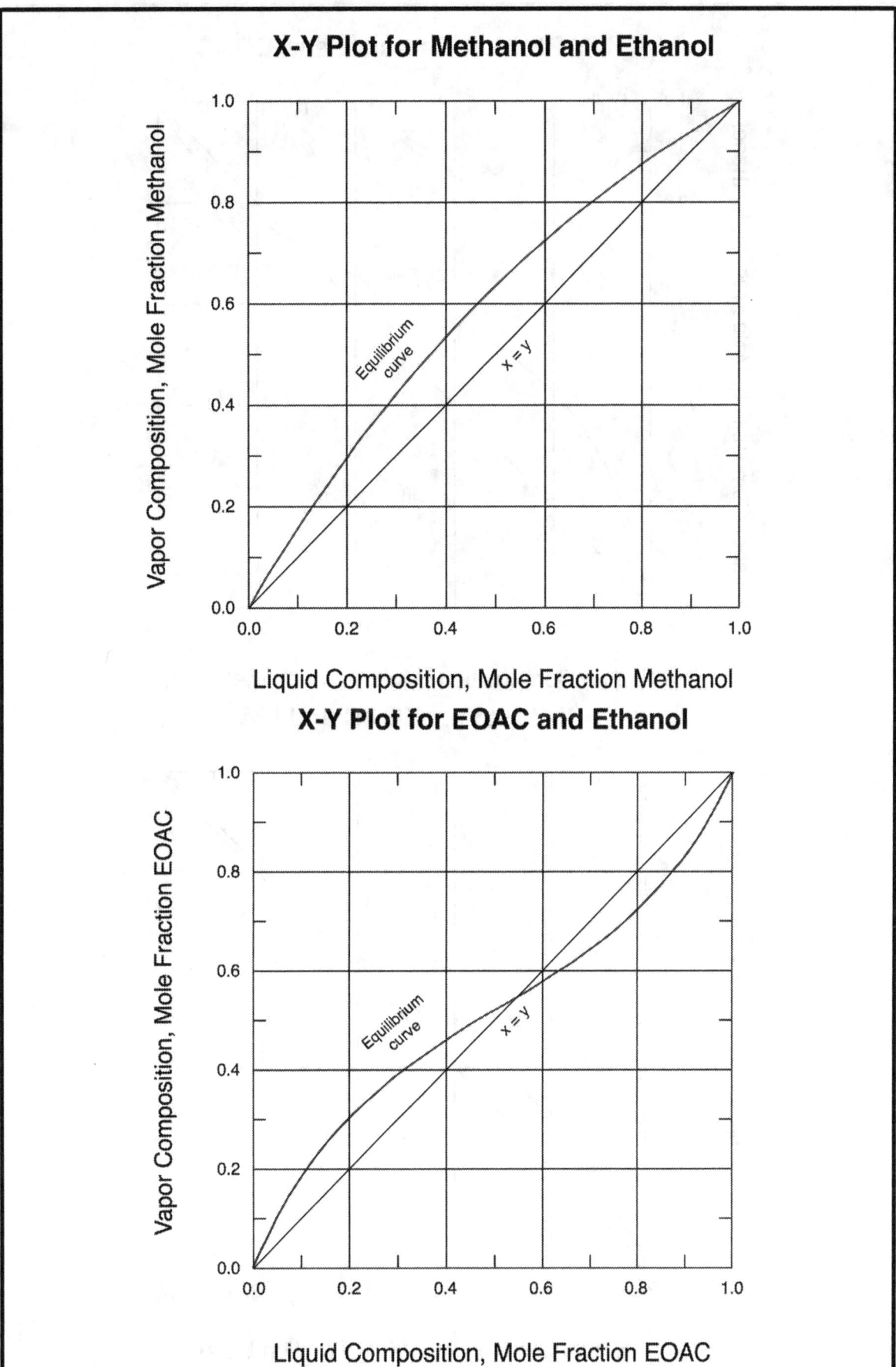

X-Y Plot for Methanol and Ethanol

Vapor Composition, Mole Fraction Methanol

Equilibrium curve

x = y

Liquid Composition, Mole Fraction Methanol

X-Y Plot for EOAC and Ethanol

Vapor Composition, Mole Fraction EOAC

Equilibrium curve

x = y

Liquid Composition, Mole Fraction EOAC

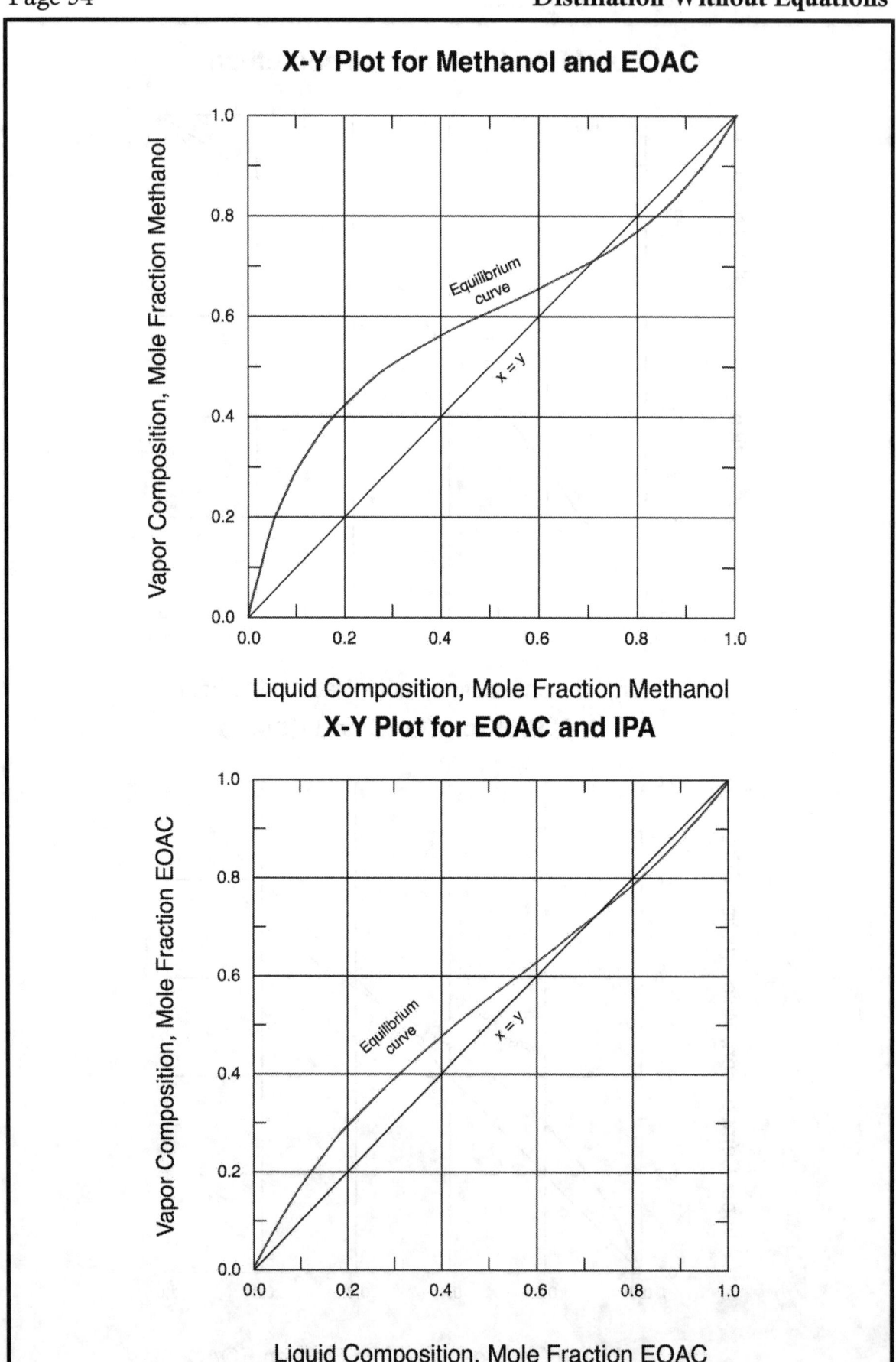

X-Y Plot for Methanol and EOAC

Liquid Composition, Mole Fraction Methanol

X-Y Plot for EOAC and IPA

Liquid Composition, Mole Fraction EOAC

Suitability, advantages, & limitations of VLE models

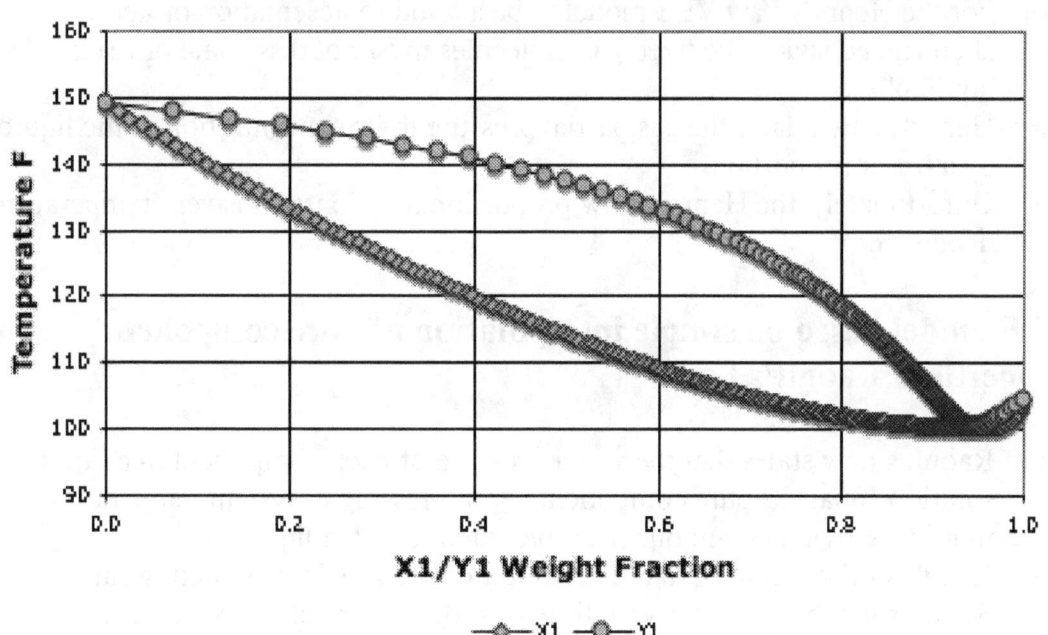

Dichloromethan / Methanol at 15.0 psia By UNIQ

Types of VLE models

- VLE models based on simple interpolation of pure component properties:
  - Henry's law, Raoult's law, classical "Steam Distillation"
- VLE models based on reduced pressure & temperature (T/T critical) generalized compressibility correlations:
  - RK, SRK
- VLE models based on activity coefficient concept, with actual binary VLE measurements regressed to produce "Best-fit" Binary Interaction Parameters (BIPS):
  - Wilson, Van Laar, NRTL, UNIQUAC
- VLE model based on activity coefficient concept, with molecule functional group contribution sums (regressed from many real binary VLE pairs) used to calculate predicted (not directly measured or regressed) BIPS:
  - UNIFAC

VLE model based on a linear gas solubility model & pure liquid component properties: Henry's Law

- Henry's Law is useful for modeling fixed gases dissolved in liquids
- For the Henry's Law VLE model to be a good representation of actual chemical behavior, the fixed gas molecules must not dissociate or react in the liquid phase
- Henry's Law relates the gas partial pressure to its concentration in the liquid by a linear proportion
- Unfortunately, the Henry's Law proportionality constant is very temperature dependent

VLE model based on simple interpolation of pure component properties: Raoult's Law

- Raoult's Law states that the vapor pressure of each component in a liquid solution equals its pure component vapor pressure at system temperature, multiplied by that component's mole fraction in the liquid phase
- Raoult's Law is similar to the ideal liquid solution density mixing rule
- Systems in which there is very little gain or loss of volume when pure component liquids are mixed, are the same systems for which the mixture VLE is likely to be well modeled by Raoult's Law
- Pure hydrocarbons of many types, away from their critical temperatures, & liquefied inert gases, can be accurately modeled by Raoult's Law

VLE model based on summation of two liquid phase vapor pressures: "Steam Distillation"

- Classical "Steam Distillation" accurately models an ancient process by which fragrances & essential oils were (& still are) distilled from plant materials
- Steam distillation processes involve boiling water & another poorly water miscible compound, together at the same time
- In the steam distillation VLE model, the water & the other compound exhibit their own vapor pressure at the system temperature
- Each of the two act as if the other's vapor was not even there:
  - Each acts as if it were being distilled, alone, at reduced system pressure
- This model is useful in many petroleum refining & hydrocarbon applications where water is present

Non-ideal VLE models based on pure component compressibility

- VLE models based on reduced pressure & temperature (T / T critical) generalized compressibility correlations were developed primarily for petroleum & petrochemical uses
- Most lab measurements & development work date from 1930 through 1970
- Redlich-Kwong (RK) & Soave-Redlich-Kwong (SRK) correlations, in original & various "extended" versions, survive & continue to be used today, working well for hydrocarbons

Activity coefficient concept applies correction factors to Raoult's Law partial pressures

- Activity coefficients are useful for calculation purposes for systems which do not closely follow one of the VLE models that are based on pure component properties
- Activity coefficients are multiplied times the partial pressure as computed by Raoult's law, to yield a corrected partial pressure
- Correction factors or activity coefficients frequently range between 0.1 & 100, with activity coefficients over 1000 common in partially miscible liquid systems
- Attempts have been made to relate activity coefficient models to molecular interaction forces, but no unifying theory seems to explain non-ideal VLE across a broad range of system types
- As a consequence, there are activity coefficient models which reproduce the VLE behavior of certain systems very well, & other systems very poorly

VLE models based on activity coefficient concept

- Usually the best result in modeling binary or multi-component VLE will occur when actual binary VLE measurements have been regressed to produce "Best-fit" Binary Interaction Parameters (BIPS) for one of the common activity coefficient models
- Models of this type in common use in 1975 included Wilson & Van Laar, whereas the NRTL & UNIQUAC models are most often used today, making their library BIPS easier to find, but not making their accuracy necessarily better
- Some of these models follow the actual VLE curves of certain binary pairs better than others, while some fail to model XY behavior well

VLE models based on activity coefficient concept

- Always generate XY VLE from your specific BIPS & VLE model, & compare to the best measured data you can find, before judging the accuracy of your model
- Do not ignore binary pairings for which there are no library BIPS available
  - Some method must be used to fill in missing BIPS, or very misleading, false, fictional results can arise during computer simulation

Another type of VLE model based on the activity coefficient concept

- The UNIFAC model does not use library BIPS for each binary pair of components, as do the activity coefficient models discussed previously
- Instead, the UNIFAC model examines the structure of each molecule, with molecular functional group contributions (regressed based on many real binary VLE pairs) summed for each molecule's structure
- UNIFAC uses the group contribution sums to calculate predicted (not directly measured or regressed) BIPS that can be used in other activity coefficient based VLE models
- UNIFAC is commonly used by process simulators to generate predicted BIPS, where library BIPS are lacking for some component pairs
- UNIFAC, as a predictor, can lead to false results, so always generate XY diagrams or do lab work to evaluate them
- Try not to start relying on UNIFAC so much as to overlook using actual binary data in binary activity coefficient models

Vapor Liquid Liquid Equilibrium (VLLE) models

- One vapor - two liquid phase equilibrium (VLLE) models exist for NRTL, UNIQUAC, & other activity coefficient correlations which are of a form to permit two liquid phases to exist
- These models can work very well when their interaction parameters have been regressed from carefully measured VLLE data for the chemical components in question
- VLLE models applied using BIPS derived from VLE data alone, can give results which are very wrong, yet comfortably misleading
- Never overlook making actual liquid-liquid partitioning measurements on which to base VLLE process development

Thermodynamic restrictions to binary VLE models

- Thermodynamic theory leads to many restrictions on the possible VLE behavior of mixtures
- People have developed many thermodynamic consistency tests which can be applied to determine whether VLE data or VLE correlation outputs are thermodynamically consistent
- Warning No. 1: Activity coefficient data or XY models produced by computations using any of the VLE models will nearly always pass the thermodynamic consistency test, no matter how far it deviates from actual system behavior
- Warning No. 2: Maxwell, Lord Kelvin, & other great thinkers were not consulted when Nature determined how chemicals would really interact, so 100% genuine & highly accurate VLE measurements can, & frequently do, fail thermodynamic consistency tests
- As far as I know, Nature has no plans to start conforming to man-made restrictions on VLE behavior, so we must not reject carefully measured data that does not conform to our theories

VLE models for binary pair behavior, plus material & energy balancing, determine the compositions calculated by a computer process simulator

Dichloromethan / Methanol at 15.0 psia By UNIQ

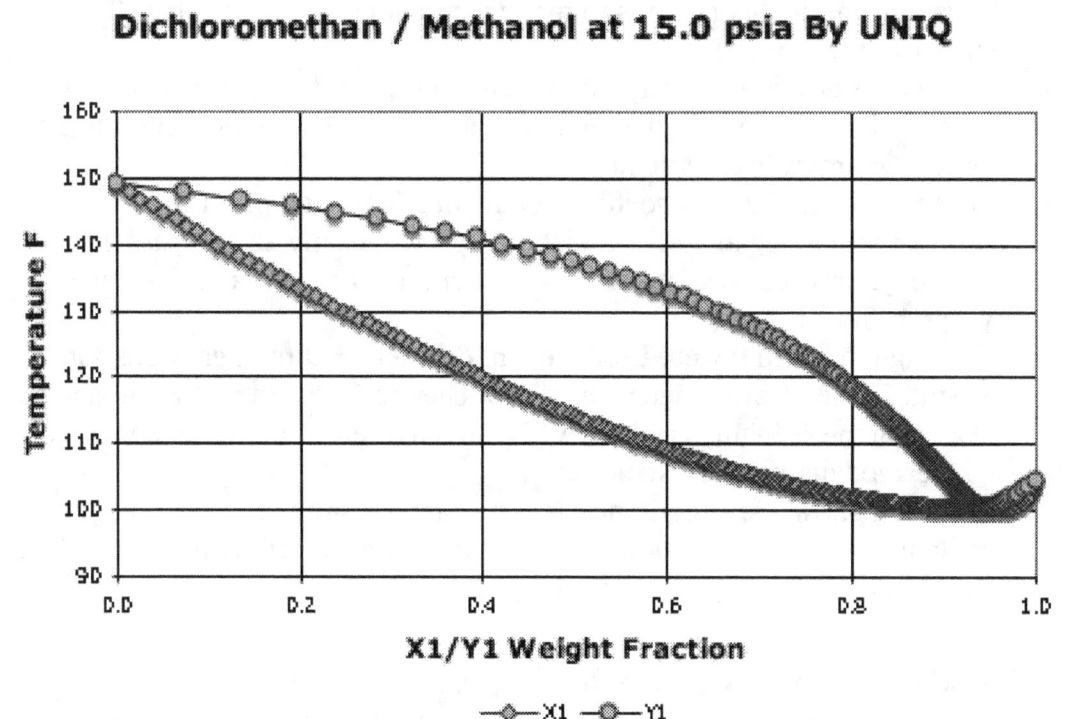

Dichloromethane / Methanol at 15 psia By UNIQ

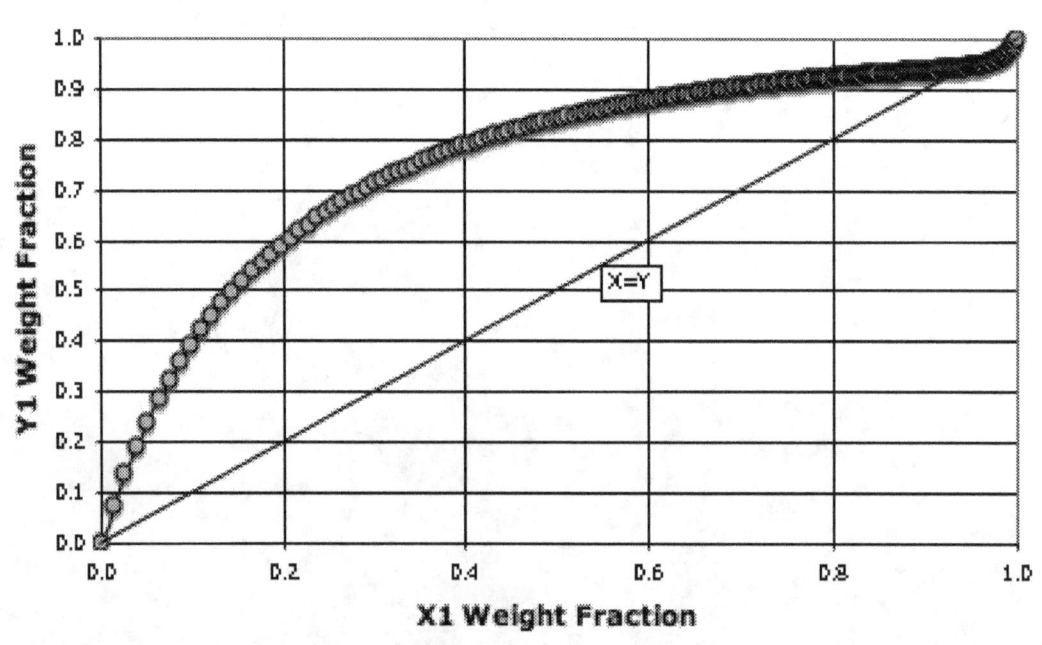

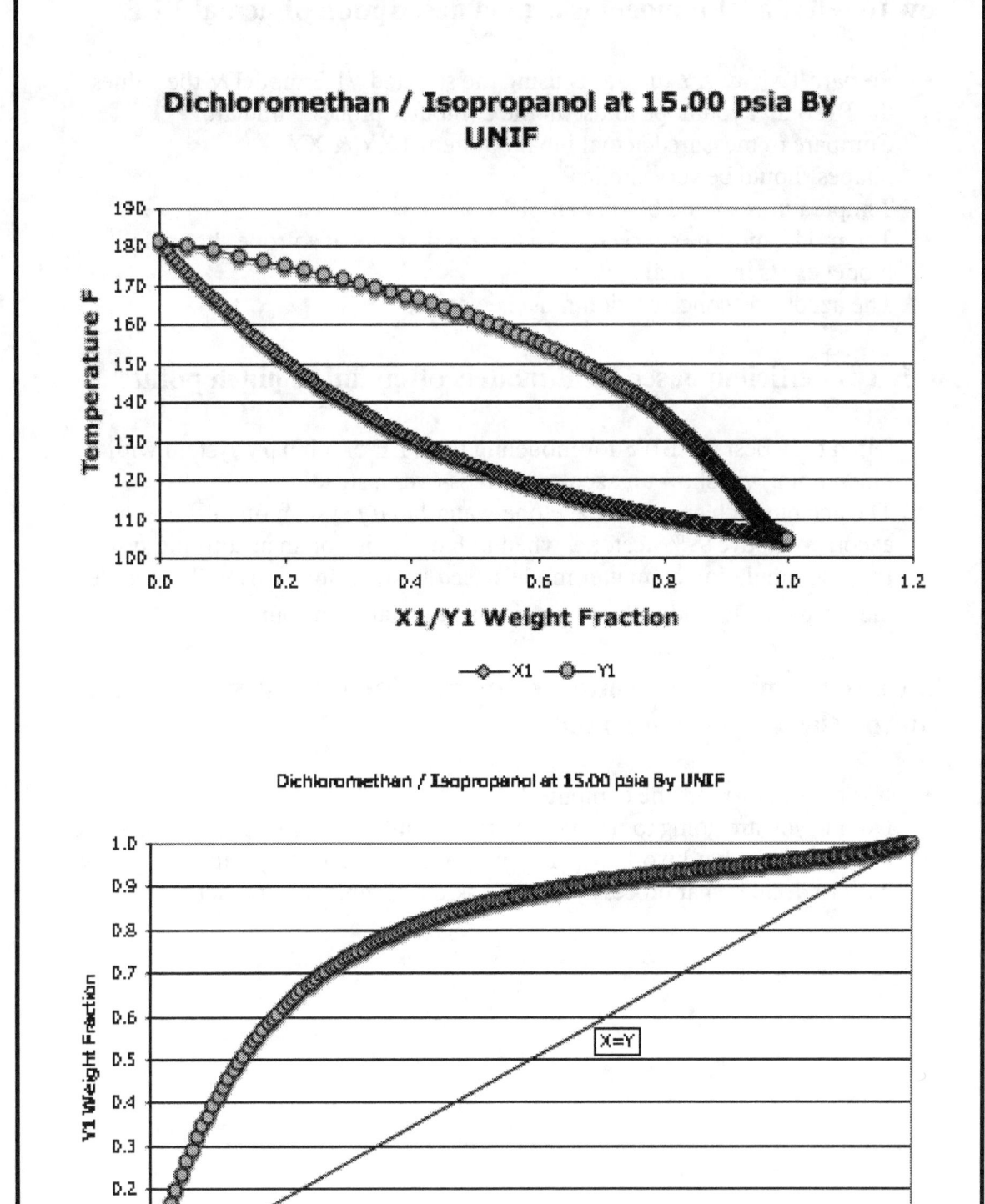

Dichloromethan / Isopropanol at 15.00 psia By UNIF

How to tell if a VLE model is a good description of actual VLE

- Prepare TXY & XY diagrams using the selected VLE model & the values of the BIPS that you hope to use in the computer process simulator
- Compare to measured actual binary system TXY & XY
- Shapes should be very similar
- Temperatures should be very similar
- The model must be checked to be certain that any azeotrope shown in the model exists in the real system
- The azeotrope concentration must be correct

Activity coefficient based VLE models often fail at pinch points

- Often the "Best fit" BIPS for modeling the VLE of a binary system with a pinch point will show an azeotrope cross-over instead!
- The acetone-rich end of the acetone-water binary system often shows an azeotrope above 99% acetone, when in fact this is not an azeotropic system
- Process simulation computer models used in such situations will fail, unless the binary VLE is adjusted to get rid of fictional azeotropes

How to determine what variety of distillation process will "un-mix" & purify the desired compound

- It is time to turn off the computer!
- Even if you are going to use a computer to simulate the distillation process, & let it do all the hard work of performing material & energy balances, first you have to decide what process you will ask the computer to model

Determine what goes up & what goes down

- List the boiling temperatures of all components & all azeotropes existing among those components, in ascending order, coolest temperature at the top
- The azeotropes can be thought of as behaving like pure components in this evaluation
- This list will tell you what goes up the most easily
- Look at the list to find a cut point, where you can let everything above that point go up, & the rest go down
- If you need to collect the second most volatile material on the list, consider a partial condenser, & consider a liquid side draw near the top of the column
- If you need to collect or remove a material mid-way through the list, consider a liquid side draw & possibly a second distillation column
- If you need to collect the second least volatile material on the list, consider a vapor or liquid side draw near the bottom of the column
- If the least volatile material is your product, collect it from the column bottom

Determine the overhead condensate condition

- The overhead condensate may form one liquid phase
- The overhead condensate may form two liquid phases, one usually being water
- There can be two organic liquid phases or two organic liquids plus water - This does happen!
- Determine the specific gravity of the phases
- If the overhead condensate forms two or three liquid phases, decide which phase the reflux must come from, heavy or light
- Determine which phases represent product or waste streams

Determine if side draws are required - liquid or vapor

- Determine if side draws are required to remove mid-range components from the column
- Determine whether each side draw is better taken as a liquid or a vapor
- Determine relatively how far up on the column each side draw should be
- Determine whether the main distillation column should be split into two columns to permit mid-range component recovery, rather than side draw

Determine if pinch points will make purification impractical

- Look at the VLE diagrams closely to determine if pinch points appear
- If pinch points are found, investigate deeply to be certain that they are pinches rather than azeotropes
- Evaluate the L/V ratio required to make progress purifying the key components in the pinch zones
- Look at the boil up heat & column size required to purify in the pinch zones
- Evaluate the economics of recovering the pinched product to decide whether it is worthwhile

Consider thermal decomposition

- Look at the thermal stability of all of the compounds present in the column
- If any compound is suspected of undergoing thermal decomposition at the column's operating temperature, investigate what temperature would be safe & consider lowering the column operating pressure to achieve a safe maximum exposure temperature
- Investigate thermal stability improvement through the removal of destabilizers or addition of stabilizers

Beware of ionic compounds & reactive component pairing

- While ionic compounds can be distilled in their neutral non-ionic forms, care is required to make certain that the ionic equilibrium drives the majority of the compound into its neutral form
- In the case of mixtures containing ammonia or amines, the pH must be high
- In the case of HCl & other acid gases, the pH must be low
- Heat of solution greatly affects the energy requirement for distillation of ionic compounds

Distillation of reactive compounds

- Consideration of reaction equilibria & reaction rates are required to correctly design a distillation system which includes reactive compounds
- If reactive compounds are present but it is not desired that they be permitted to react within the distillation system, then the operating pressure & temperature may be lowered until the reaction potential is small, or reaction inhibitors may be added

Accommodate fixed gases

- If any fixed gases are present in the feed material, their distillation equilibria & reactivity potential must be assessed
- Inert gases are usually present, dissolved in most distillation feed materials - nitrogen, methane, etc.
- The system pressure control & venting system must be prepared to release gases which are dissolved in the feed, but do not significantly re-dissolve in the liquid distillation products

Determine if pump around loops are desired

- Pump around loops may be included in a distillation system design
- Usually these loops provide cooling at a point different than the column condenser, either to condense vapor to permit liquid product withdrawal in large quantities, or to balance the vapor flow rates in different sections of the column, relative to the physical capability of the installed distillation trays or packing

Determine if extractive distillation or entrainer distillation would be useful

- If your examination of the component VLE does not point towards a likely candidate distillation process, consider extractive or entrainer distillation
- Extractive distillation pumps an additional, less volatile, liquid feed stream into a distillation column, above the primary feed inlet
- Its effect is to reduce the relative volatility of one or more components that otherwise would have gone overhead with desirable product
- Entrainer distillation traps a large physical inventory of an additional light component in the decanter, reflux system, & the upper part of the column, where it enhances the volatility of one or more compounds that otherwise would go down the column into the bottoms product
- Both extractive & entrainer distillation typically require that a second column be provided for regeneration

Make initial educated guesses about reflux ratio, heat input, & number of distillation stages

- Look at one of the key component pair distillation XY diagrams
- Construct a tried-&-true McCabe-Theile diagram on that XY diagram
- Use the McCabe Theile diagram to determine minimum L/V above & below the feed point
- On a McCabe-Theile diagram constructed with a reasonable reflux ratio, 1.5 times the minimum reflux, use graphical construction to determine the approximate number of theoretical stages

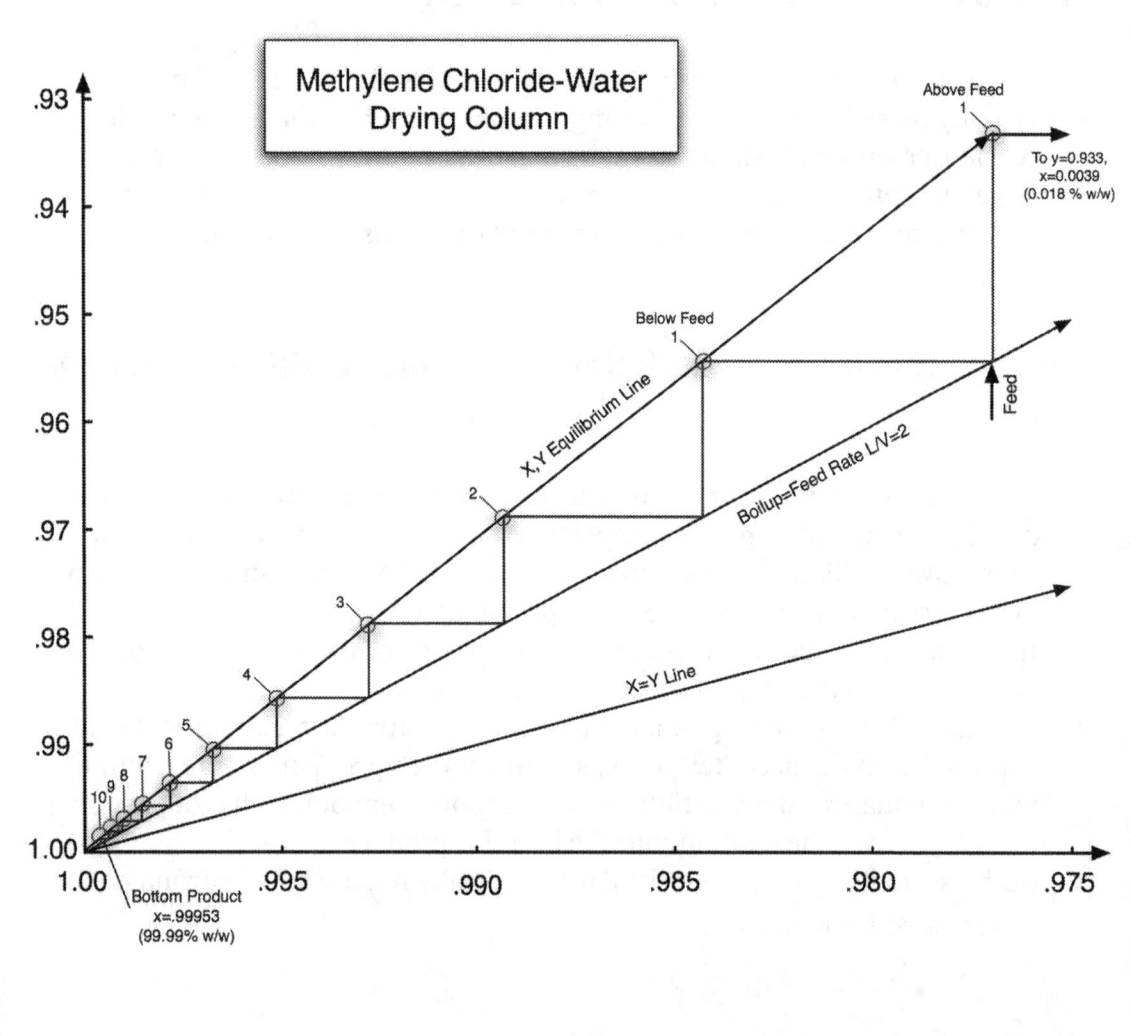

How to use a computer process simulator to model a candidate distillation process

- Set up topographical flow sheet model
- Select components from library, or create new components or pseudo components
- Select VLE correlation options
- Retrieve VLE BIPS from library
- Evaluate TPXY relationships calculated from the selected VLE model
- Adjust BIPS or in-fill using UNIFAC until all binary pair TPXY are close to the truth
- Input temperature, pressure, state, & quantity data for feed streams, cut streams, side draws

Set up model constraints & input equipment & convergence specifications

- Define equipment:
    - Flash blocks for setting feed stream conditions, mixers where multiple feeds go to one column, feed preheaters, product coolers, interchangers
- Define distillation column fixed parameters:
    - Number of stages, feed & product locations, condenser & reboiler type, column pressure profile
- Define distillation column calculation constraints:
    - One constraint will determine the column bottom condition (use fixed reboiler heat input for the first trial run) & another constraint will determine the column top condition (use fixed reflux ratio for the first trial run)
- Set convergence criteria & damping factors

Provide estimated distillation column internal flow & temperature profiles

- Column flow & temperature profiles must be provided, for the computer simulator to have a starting point for its calculations
- Usually it is sufficient to input a reasonable top & bottom temperature for the column, & reasonable top & bottom molar flows, telling the simulator program to interpolate & fill in intermediate stage data
- After one converged run has been made on the system you are studying, successive runs usually will converge more readily if the column profile computed during a previous run is used as the starting point, rather than the coarsely interpolated initial estimated profile

Run initial simulation, adjusting input parameters until one run converges with reasonable results

- If your first set of input parameters runs to convergence with reasonable results (rather than zero overhead vapor flow or other correct but trivial converged solutions) you are very lucky:
  - You input a good guess for the initial tower profile
- If your first run does not converge, heed the warnings given in the simulator output
- It is very easy to under estimate the heat input required, or to broadly miss the correct column flows, on the high or low side, so change the input & try again
- It is also very easy to mistake the engineering units used by the simulator model for other similar units, so when the process simulator results seem to be off by a large or familiar factor (like 1.8, 2.2, 454, or 1000), check the units, reinput, & try again

After one simulation has converged: Evaluate, adjust, rerun

- After one simulation has converged, evaluate the output product streams quantity & quality, & energy used
- Adjust the primary column input parameters to improve deficiencies in quantity or quality:
  - Move the feed point up or down, increase the number of theoretical stages
- Increase the reboiler heat input & increase the reflux ratio to improve product purity, but reduce the heat input & the reflux ratio to decrease hot & cold utility usage
- Do not be afraid to make many variations & run the simulator over & over again, to see what happens if this or that increases or decreases
- Limit the size of step changes made to input parameters, to increase the chance of new runs converging successfully

The opportunity window for optimization opens now, & closes when the final simulation run becomes the design basis for column, internals, & overall distillation system design

Make use of the process optimization opportunity now

- Now that the simulation gives close to the desired result, it is time to run more simulations in a parametric block
- One parameter to reduce is the number of theoretical stages - try 10%, 20%, 50%, 80% fewer!
- Adjust the feed point because the result may surprise you:
  - Try moving the feed up & down because either may yield improved purity
- Step all side draws up & down in position & in flow rate

Make use of the process optimization opportunity now

- In the optimization runs, permit the more advanced column top & bottom convergence specification criteria their chance to decrease energy input or reflux quantity:
  - Try using purity specifications or flow quantity specifications instead of heat input & reflux ratio as in the initial runs
- Run column tray or packing sizing software on these optimization run outputs to provide physical size input for cost estimation
- The optimal design will balance fixed & variable life cycle costs

Column internal selection:
Which trays or which packing?

- Unless low pressure drop (< 0.1 psi per stage) is necessary, for column diameter > 2 feet (500 mm), consider valve trays first
- If pressure drop of near 0.01 psi per stage is required, in non-aqueous low pressure systems, consider structured packing first
- For low pressure drop in aqueous systems, consider random packing first at or above atmospheric pressure, consider structured packing first under vacuum
- For high vacuum conditions (1 to 20 mmHg absolute) consider woven wire mesh packing first, knit wire packing second
- Valves for valve trays are available round or rectangular, with directional patterns such as checkered, basket weave:
  - All will work, with variations in cost & maximum & minimum rate at a given diameter & tray spacing key among selection criteria
- Floating valves act as check valves on the tray deck, reducing liquid weeping at low vapor rates, improving performance during turn down
- Punched raised louvers, also called fixed valves, resist fouling & are mechanically very strong, good when turn down is not important

Rate the physical distillation column just selected for alternative operations

- Often a distillation system will be used for more than one process
- After simulating the selected primary process & optimizing the physical distillation column for that process it is time to simulate alternative processes
- If the required alternative processes can be performed in the physical system selected for the primary process no design modifications will be needed, but their process information should appear in the design basis because it is not certain which system will demand the highest duty or dictate maximum size of the reboiler, condenser, coolers, pumps, instruments, or auxiliaries
- If the alternative process requires more separation stages, a different feed point, different decanter or reflux provisions, or much higher or lower energy input, incorporate this information into the design basis specification, in addition to the complete primary design

Example of process design for a multi-purpose distillation system

- The following pages portray the computer simulation output from rating a newly designed distillation system to process five different feed streams on a campaign basis
- The column & all of its auxiliaries were built with all features needed to handle all of these streams
- The maximum design case for the column was for the most difficult fractionation:
  - Acetone dehydration
- The reboiler design considered two maxima:
  - Peak heat input (thus peak heat flux) during methanol recovery, & lowest available temperature difference during toluene recovery
- The condenser design considered two maxima:
  - Peak cooling demand with methanol, & lowest temperature difference during methylene chloride recovery

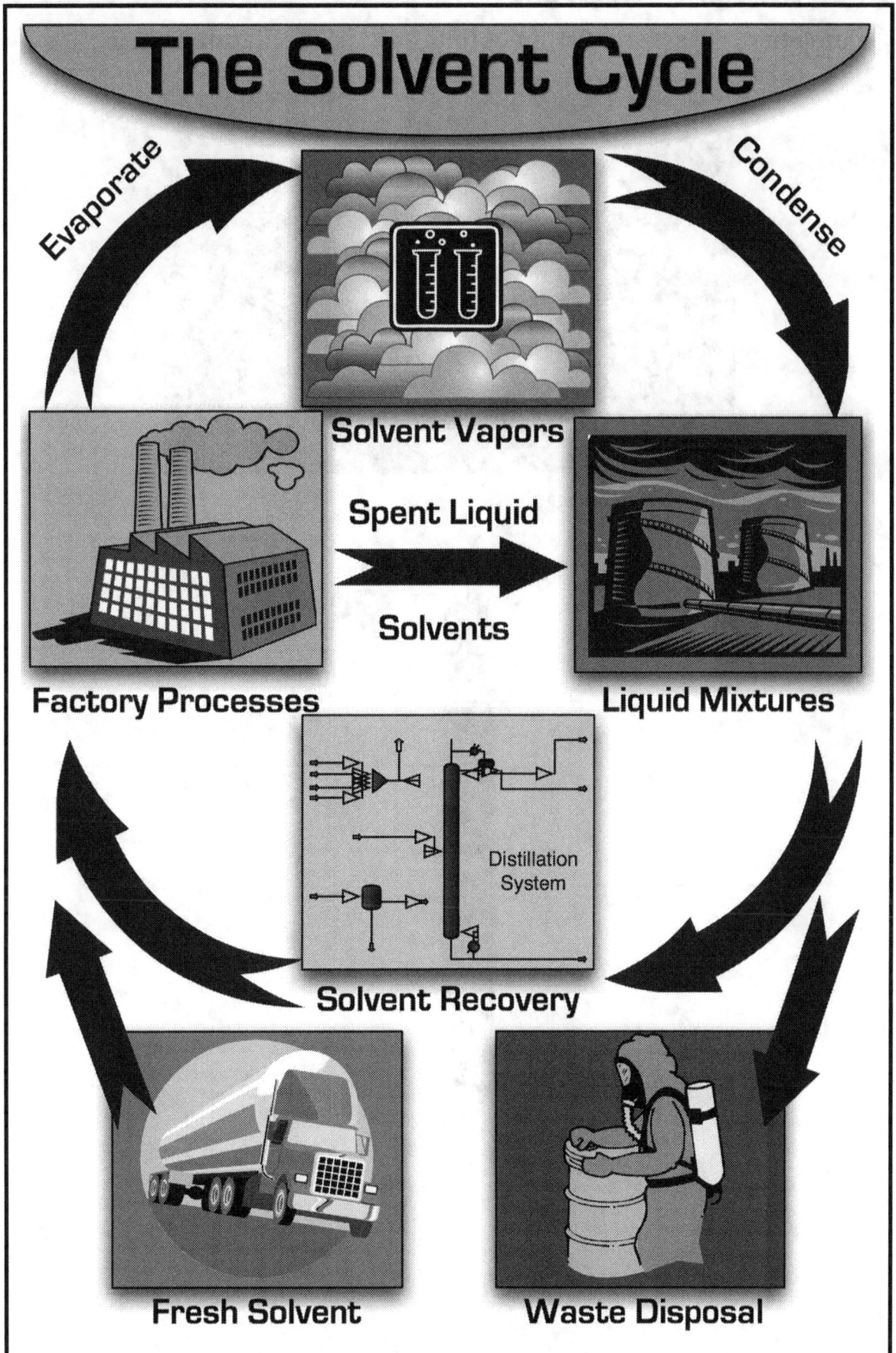

The Solvent Cycle

Evaporate

Condense

Solvent Vapors

Spent Liquid Solvents

Factory Processes

Liquid Mixtures

Distillation System

Solvent Recovery

Fresh Solvent

Waste Disposal

Computer process simulation output for this distillation system

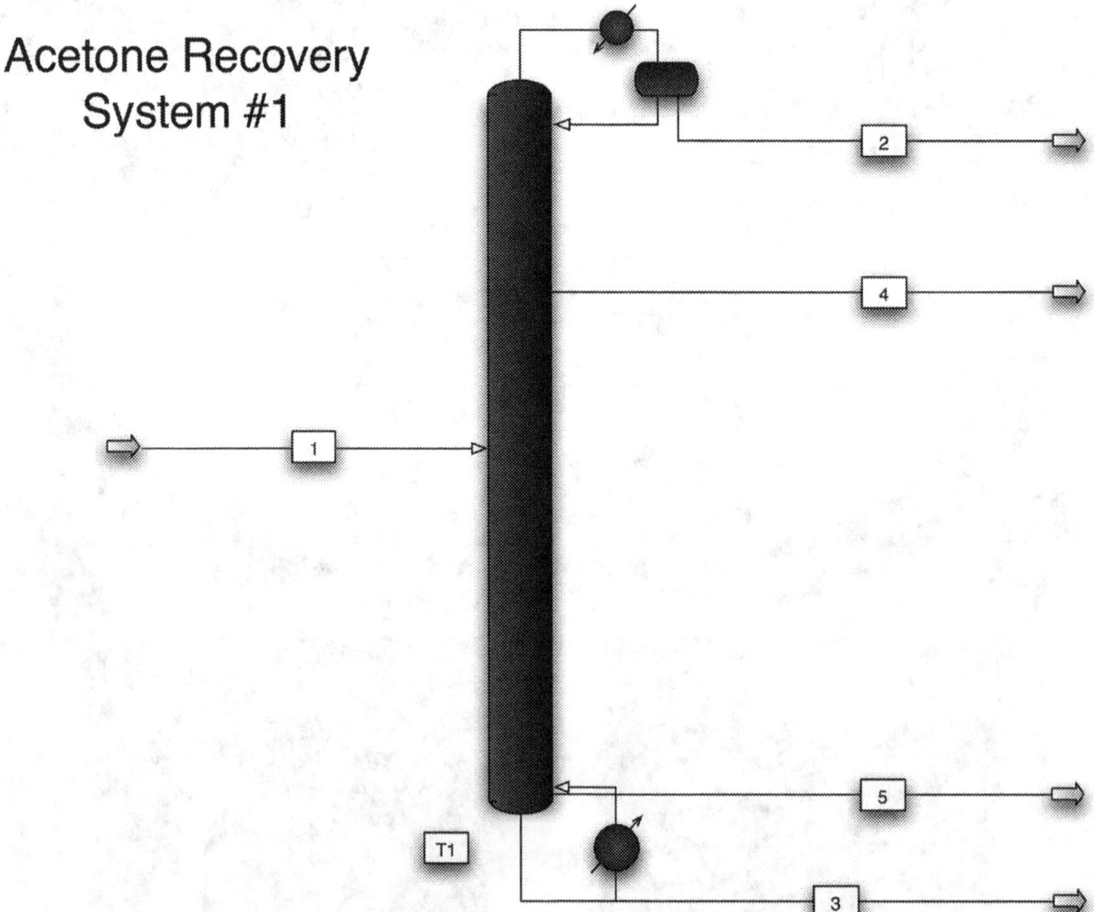

Acetone Recovery
System #1

1

2

4

5

3

T1

Acetone recovery material balance: Distillate product

Steam Name	1	2	3	4	5
Stream Description	Feed	Main Product	Vapor Side	Liquid Side	Bottoms
Phase	Liquid	Liquid	Liquid	Vapor	Liquid
KG/HR	2800.000	2600.000	0.022	0.029	199.967
Temperature C	10.000	54.814	81.248	57.789	81.248
Pressure BAR	4.000	1.000	1.320	1.067	1.320
Molecular Weight	51.064	57.558	43.730	57.047	20.700
Weight Comp. Percent					
H2O	6.1780	0.4078	14.7548	0.814	81.1975
Acetone	93.8220	99.5922	85.2452	99.1856	18.8025
Weight Comp. Rates KG/HR					
H2O	172.9830	10.6037	0.0186	0.0002	162.3680
Acetone	2627.0171	2589.3970	0.0032	0.0283	37.5988
Enthalphy M*KCAL/HR	0.025	0.077	0.0000	0.0000	0.015

Acetone recovery column profile, T-P-L-V-Q

			Column Summery				
			Net Flow Rates				
Tray	Temperature	Pressure	Liquid	Vapor	Feed	Product	Duties
	C	BAR	KG-MOL/HR				K*KCAL/HR
1	54.8	1.00	189.7			45.2	-1.6886
2	56.4	1.02	190.6	234.9			
3	56.9	1.04	190.5	235.7			
4	57.3	1.05	190.4	235.7			
5	57.8	1.07	190.4	235.6		0.0	
6	58.2	1.08	190.3	235.5			
7	58.7	1.10	190.2	235.5			
8	59.1	1.11	190.1	235.4			
9	59.6	1.13	189.9	235.2			
10	60.0	1.15	189.7	235.1			
11	60.5	1.16	189.3	234.8			
12	61.0	1.18	188.9	234.5			
13	61.5	1.19	188.2	234.1			
14	62.0	1.21	254.7	233.4	54.8		
15	62.4	1.22	254.7	235.0			
16	62.8	1.24	254.5	245.0			
17	63.3	1.26	254.0	244.9			
18	63.8	1.27	252.6	244.4			
19	64.6	1.29	248.1	243.0			
20	66.8	1.30	227.1	238.5			
21	81.2	1.32		217.5		9.7	1.7558

Acetone recovery column temperature profile
Acetone recovery column temperature graph, top to bottom

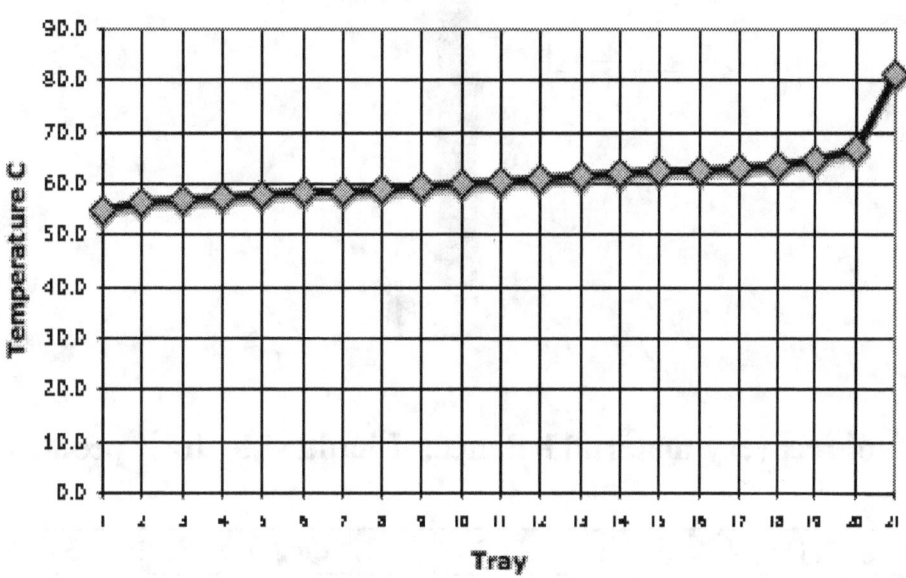

Acetone recovery column liquid & vapor flows, tray by tray
Acetone recovery column liquid & vapor flow graph

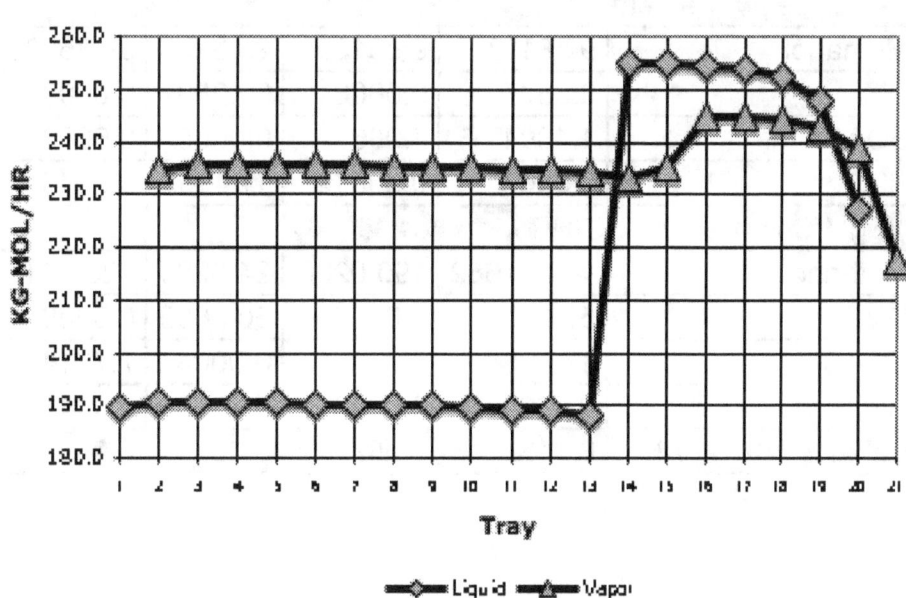

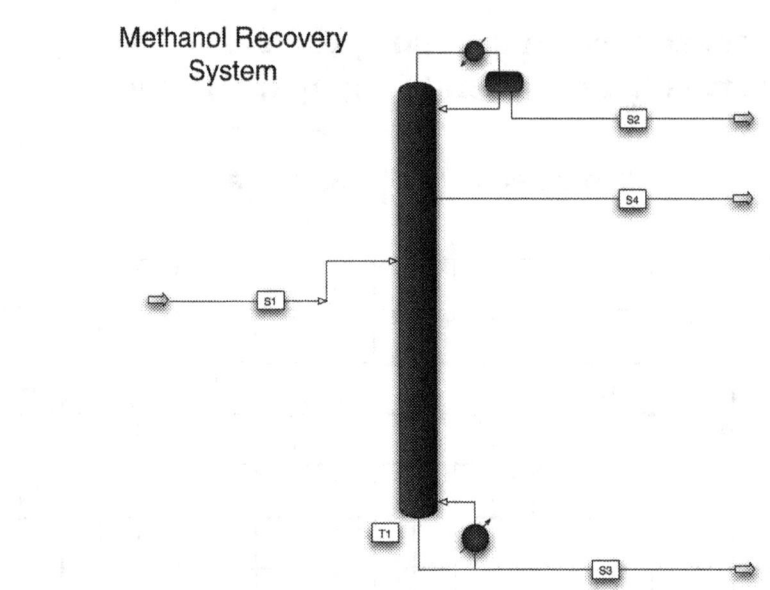

Methanol Recovery
System

Methanol recovery material balance: Liquid side draw product

Stream Name	S1	S2	S3	S4
Stream Description	Feed	Waste EOAC	Bottom Waste	Main Product
Phase	Liquid	Liquid	Liquid	Liquid
KG/HR	1500.000	200.000	85.000	1215.000
Temperature C	20.000	64.118	87.677	65.216
Pressure BAR	1.100	1.013	1.213	1.045
Molecular Wight	32.740	33.092	42.378	32.171
Weight Comp. Percents				
Methanol	94.8112	95.0130	29.3276	99.3592
DMF	3.3577	0.0000	59.2537	0.0000
EOAC	1.1804	4.9869	0.0005	0.6363
H2O	0.6507	0.0001	11.4183	0.0045
Weight Comp. Rates KG/HR				
Methanol	1422.1682	190.0260	24.9284	1207.2139
DMF	50.3656	0.0000	50.3656	0.0000
EOAC	17.7054	9.9737	0.0004	7.7314
H2O	9.7604	0.0003	9.7055	0.0547
Enthalpy M*KCAL/HR	0.013	0.007	0.006	0.044

Methanol recovery column profile, T-P-L-V-Q

			Column T1 Summery				
			Net Flow Rates				
Tray	Temperature	Pressure	Liquid	Vapor	Feed	Product	Duties
	C	BAR	KG-MOL/HR				K*KCAL/HR
1	64.1	1.01	169.9			6.0	-1.5275
2	64.3	1.00	169.5	176.0			
3	64.6	1.02	169.3	175.5			
4	64.9	1.03	169.2	175.3			
5	65.2	1.04	131.5	175.3		37.8	
6	65.5	1.06	131.5	175.3			
7	65.8	1.07	131.5	175.3			
8	66.0	1.08	131.5	175.3			
9	66.3	1.09	131.5	175.3			
10	66.5	1.10	131.5	175.3			
11	66.9	1.11	182.5	175.3	45.8		
12	67.2	1.12	182.5	180.5			
13	67.4	1.13	182.6	180.5			
14	67.7	1.14	182.6	180.6			
15	67.9	1.15	182.6	180.6			
16	68.2	1.16	182.6	180.6			
17	68.6	1.17	182.4	180.6			
18	69.2	1.18	182.0	180.4			
19	70.2	1.19	180.6	179.9			
20	73.1	1.20	176.2	178.6			
21	87.7	1.21		174.2		2.0	1.5718

Methanol recovery column temperature profile
Methanol recovery column temperature graph, top to bottom

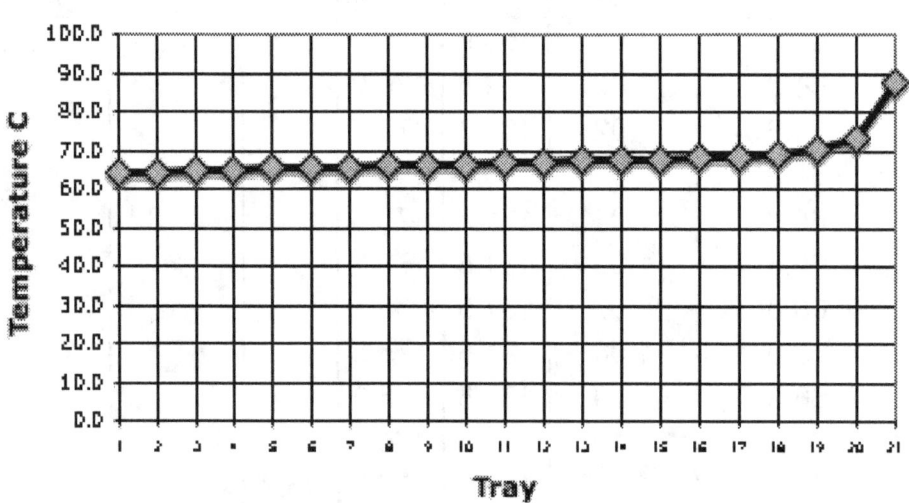

Methanol recovery liquid & vapor flows, tray by tray
Methanol recovery column liquid & vapor flow graph

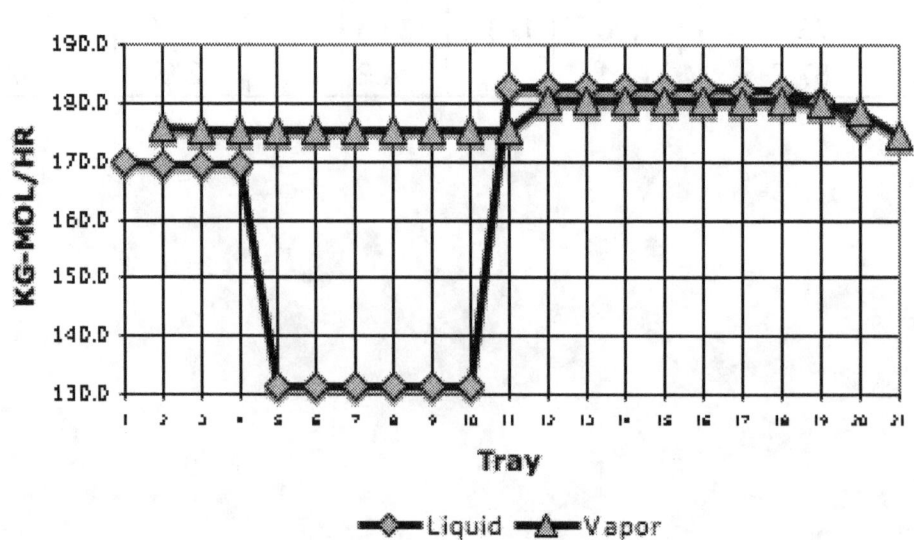

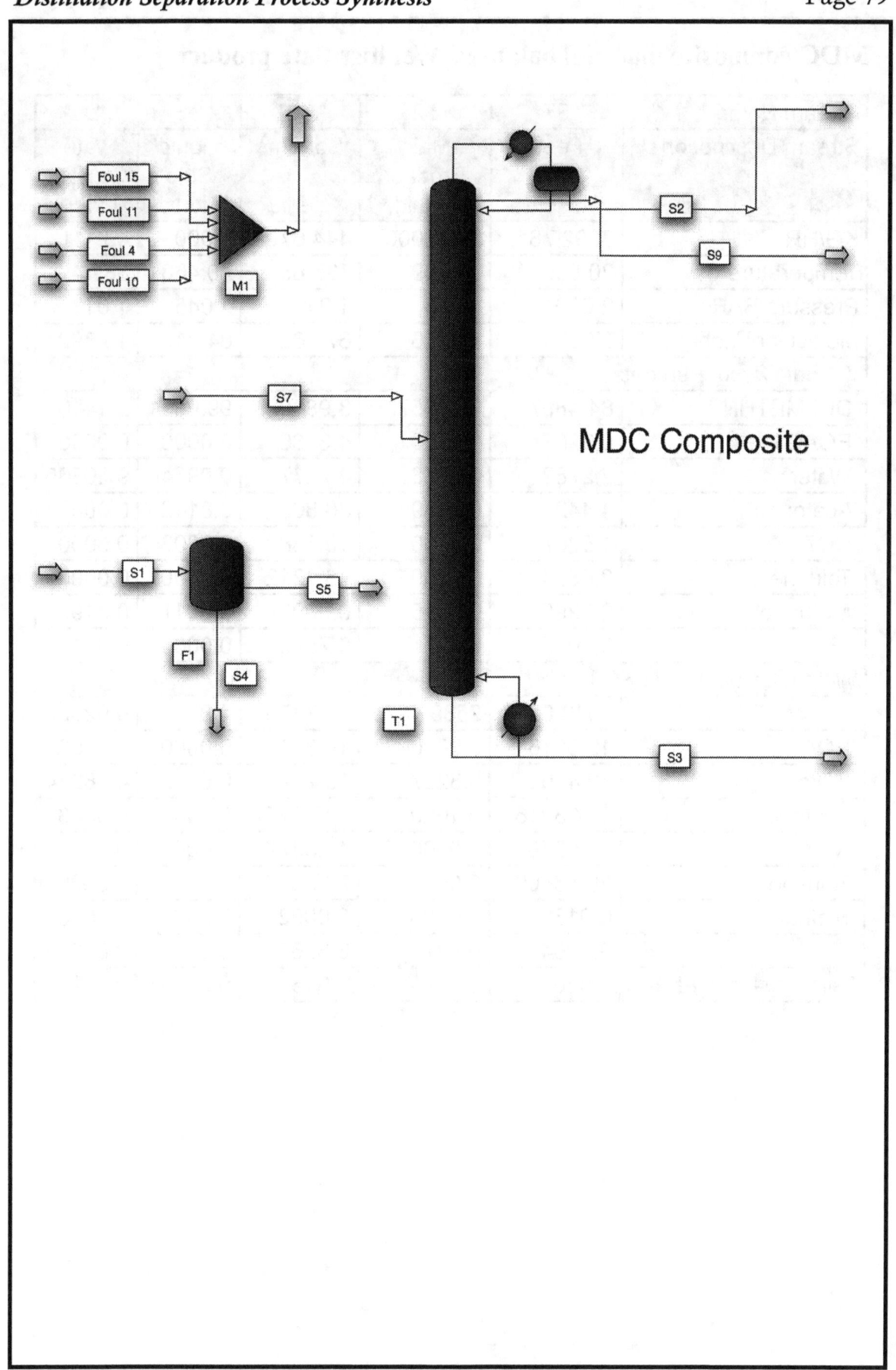

MDC composite material balance: Wet distillate product

Stream Name	S7	S2	S3	S8	S9
Stream Description	Feed	Main Product	Bottoms	Liquid Side	Water OVHDS
	Liquid	Liquid	Liquid	Liquid	Liquid
KG/HR	3092.781	2600.000	444.670	5.000	43.111
Temperature C	20.000	38.149	73.465	40.468	38.149
Pressure BAR	2.000	1.013	1.213	1.045	1.013
Molecular Wight	77.320	83.726	67.962	84.313	18.392
Weight Comp. Percents					
DCLMETHN	84.4897	99.5957	3.9840	99.0831	2.1410
EOAC	0.6267	0.0000	4.3586	0.0000	0.0000
Water	2.2782	0.3778	4.2277	0.0954	97.0369
Acetone	4.4423	0.0039	30.8652	0.8142	0.0008
NC7	1.5267	0.0000	10.6883	0.0003	0.0000
Toluene	6.4874	0.0000	45.1212	0.0000	0.0000
Methanol	0.0295	0.0215	0.0000	0.0011	0.8195
IPA	0.1095	0.0011	0.7549	0.0060	0.0018
Weight Comp. Rates KG/HR					
DCLMETHN	2613.0806	2589.4875	17.7158	4.9541	0.9230
EOAC	19.3816	0.0000	19.3815	0.0000	0.0000
Water	70.4599	9.8227	18.7992	0.0048	41.8334
Acetone	137.3916	0.1020	137.2482	0.0407	0.0003
NC7	47.5278	0.0000	47.5277	0.0000	0.0000
Toluene	200.6406	0.0000	200.6403	0.0000	0.0000
Methanol	0.9138	0.5603	0.0002	0.0001	0.3553
IPA	3.3854	0.0274	3.3569	0.0003	0.0008
Enthalpy M*KCAL/HR	0.020	0.028	0.016	0.000	0.002

MDC composite column profile, T-P-L-V-Q

			Column T1 Summery				
			Net Flow Rates				
Tray	Temperature	Pressure	Liquid	Vapor	Feed	Product	Duties
	C	BAR	KG-MOL/HR				K*KCAL/HR
1	38.1	1.01	215.0			33.4	-1.7151
2	39.1	1.01	216.7	248.4			
3	39.6	1.02	217.2	250.1			
4	40.0	1.03	217.3	250.6			
5	40.5	1.04	217.2	250.7		0.1	
6	41.3	1.06	216.7	250.6			
7	43.1	1.07	215.8	250.1			
8	45.9	1.08	214.8	249.2			
9	48.9	1.09	213.8	248.2			
10	51.5	1.10	212.6	247.2			
11	53.5	1.11	256.4	246.0	40.0		
12	55.6	1.12	256.0	249.8			
13	57.0	1.13	255.5	249.5			
14	58.0	1.14	254.9	248.9			
15	58.7	1.15	254.2	248.3			
16	59.4	1.16	253.4	247.6			
17	60.0	1.17	252.3	246.8			
18	60.6	1.18	250.4	245.7			
19	61.6	1.19	246.2	243.9			
20	64.3	1.20	234.4	239.6			
21	73.5	1.21		227.8		6.5	1.7413

MDC composite column temperature profile
MDC composite column temperature graph, top to bottom

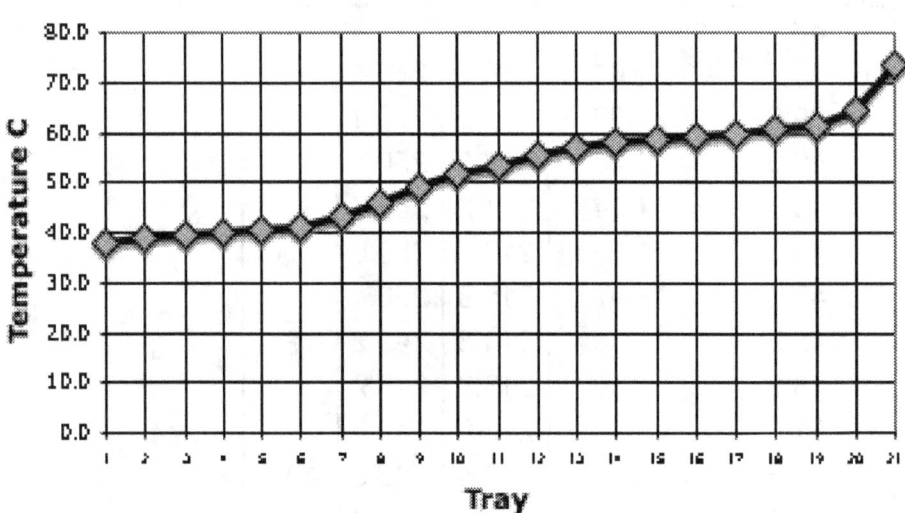

MDC composite liquid & vapor flows, tray by tray
MDC composite column liquid & vapor flow graph

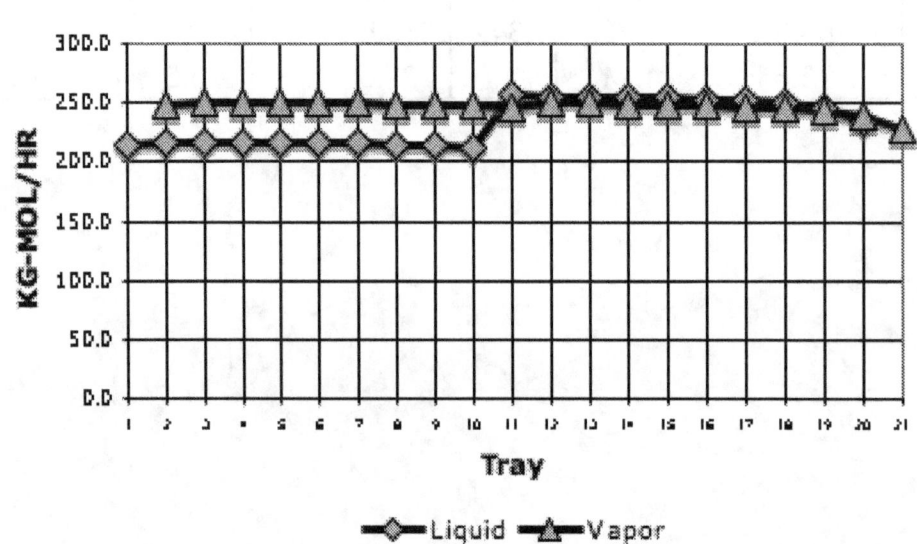

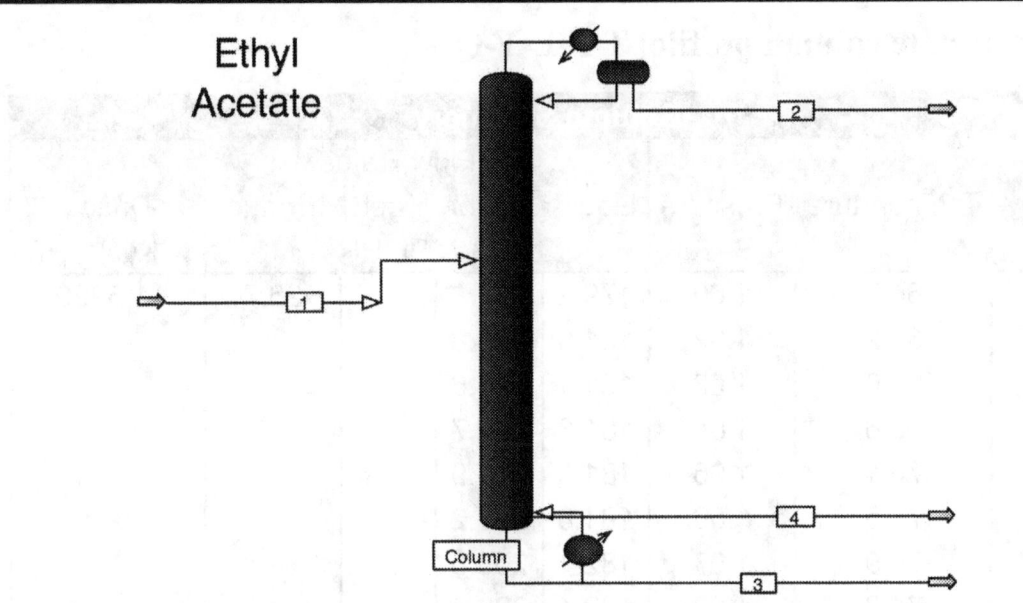

Ethyl
Acetate

Ethyl acetate material balance: Lower vapor side draw product

Stream Name	1	2	4	3
Stream Description	Feed	OVHDS Waste	Main Product	CRUD
Phase	Liquid	Liquid	Vapor	Liquid
KG/HR	2310.000	1751.617	548.383	10.000
Temperature C	10.000	64.721	82.587	82.910
Pressure BAR	4.000	1.000	1.209	1.220
Molecular Wight	69.989	65.710	87.950	88.086
Weight Comp. Percents				
H2O	4.3290	5.7033	0.0182	0.0007
Ethanol	6.2771	8.2564	0.0689	0.0093
EOAC	72.6407	63.9912	99.7706	99.9513
IPA	0.5195	0.6659	0.0611	0.0146
MEAC	16.2338	21.3832	0.0811	0.0241
Weight Comp. Rates KG/HR				
H2O	100.0000	99.9000	0.1000	0.0001
Ethanol	145.0000	144.6210	0.3780	0.0009
EOAC	1677.9999	1120.8801	547.1246	9.9951
IPA	12.0000	11.6632	0.3353	0.0015
MEAC	374.9999	374.5526	0.4450	0.0024
Enthalpy M*KCAL/HR	0.031	0.077	0.069	0.000

Ethyl acetate column profile, T-P-L-V-Q

			Net Flow Rates				Duties
Tray	Temperature	Pressure	Liquid	Vapor	Feed	Product	
	C	BAR	KG-MOL/HR				K*KCAL/HR
1	64.7	1.00	179.2			26.7	-1.5920
2	68.2	1.02	181.0	205.8			
3	69.0	1.03	181.1	207.6			
4	69.6	1.04	181.3	207.7			
5	70.1	1.05	181.5	207.9			
6	70.5	1.06	181.8	208.2			
7	70.9	1.07	182.1	208.5			
8	71.2	1.08	182.4	208.8			
9	71.5	1.09	225.2	209.1	33.0		
10	72.0	1.10	225.8	218.9			
11	72.4	1.11	226.4	219.4			
12	72.8	1.13	227.0	220.0			
13	73.1	1.14	227.7	220.7			
14	73.5	1.15	227.8	221.3			
15	74.1	1.16	225.9	221.5			
16	75.9	1.17	222.3	219.5			
17	79.2	1.18	222.8	216.0			
18	81.3	1.19	224.2	216.4			
19	83.1	1.20	224.9	217.9			
20	82.6	1.21	225.3	218.6		6.2	
21	82.9	1.22		225.1		0.1	1.7076

Column Summery

Ethyl acetate column temperature profile
Ethyl acetate column temperature graph, top to bottom

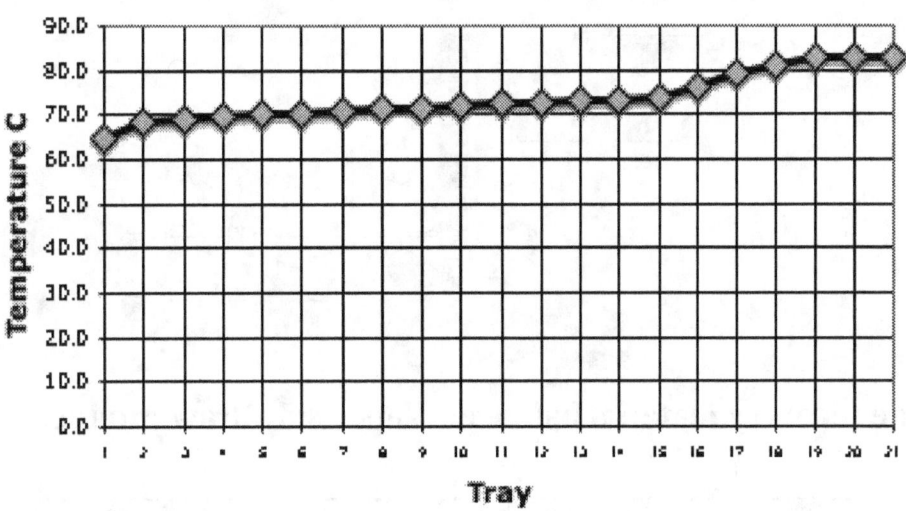

Ethyl acetate liquid & vapor flows, tray by tray
Ethyl acetate column liquid & vapor flow graph

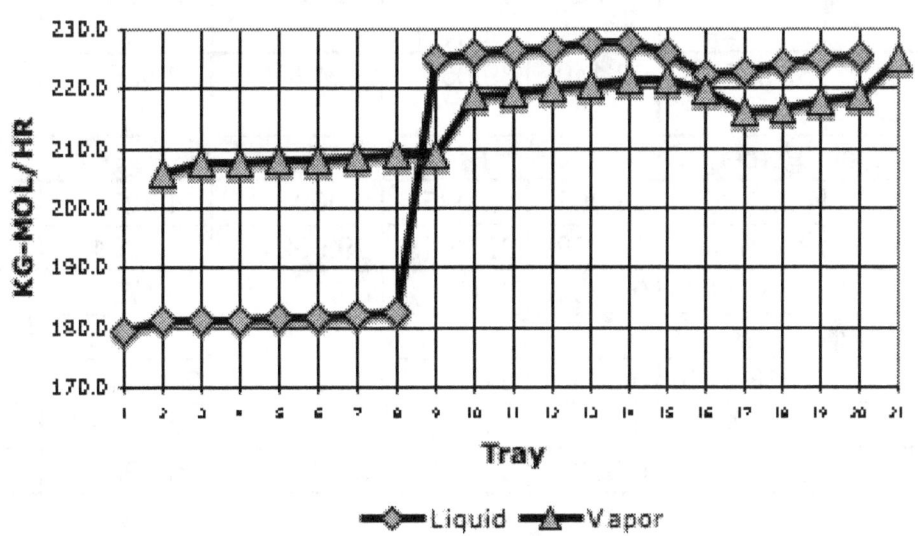

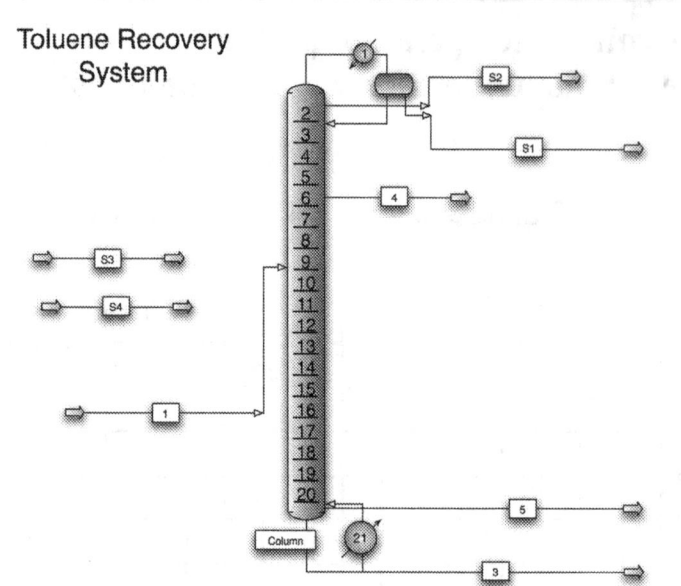

Toluene Recovery System

Toluene recovery material balance: Vapor side draw product

Stream Name	1	3	4	5	S1	S2
Stream Description	Feed	CRUD	Liquid Side	Main Product	THF Purge	Water Waste
Phase	Liquid	Liquid	Liquid	Vapor	Liquid	Liquid
KG/HR	4497.010	9.214	0.089	1375.000	2933.654	179.054
Temperature C	10.000	11.217	97.468	117.217	71.062	71.062
Pressure BAR	3.000	1.220	1.052	1.220	1.020	1.020
Molecular Wight	57.683	92.140	89.054	92.140	53.536	22.000
Weight Comp. Percents						
THF	49.7735	0.0002	12.0253	0.0006	74.8273	24.0870
Tolunene	39.0694	99.9998	87.9443	99.9994	12.7001	0.0528
Water	11.1571	0.0000	0.0303	0.0000	12.4727	75.8603
Weight Comp. Rates KG/HR						
THF	2238.3210	0.0000	0.0107	0.0087	2195.1731	43.1286
Tolunene	1756.9537	9.2140	0.0783	1374.9913	372.5757	0.0945
Water	501.7352	0.0000	0.0000	0.0000	365.9046	135.8306

Toluene recovery column profile, T-P-L-V-Q

Tray	Temperature C	Pressure BAR	Liquid	Vapor	Feed	Product	Duties K*KCAL/HR
			Net Flow Rates KG-MOL/HR				
1	71.1	1.02	186.8			62.9	-1.8778
2	65.2	1.02	184.0	249.7			
3	76.2	1.03	169.7	247.0			
4	91.1	1.04	168.4	232.6			
5	97.8	1.05	168.4	231.3		0.0	
6	100.7	1.06	168.8	231.3			
7	102.0	1.07	168.9	231.6			
8	102.7	1.08	168.9	231.7			
9	103.2	1.09	169.0	231.8			
10	103.5	1.10	257.1	231.9			
11	103.9	1.11	264.2	231.9	78.0		
12	112.1	1.13	265.0	242.1			
13	113.8	1.14	265.3	249.2			
14	114.7	1.15	265.6	249.9			
15	115.2	1.16	265.6	250.3			
16	115.6	1.17	265.8	250.6			
17	115.9	1.18	266.0	250.8			
18	116.3	1.19	266.1	250.9			
19	116.6	1.20	266.3	251.1			
20	116.9	1.21	266.5	251.3			
21	117.2	1.22		251.4		15.0	2.1186

Toluene recovery column temperature profile
Toluene recovery column temperature graph, top to bottom

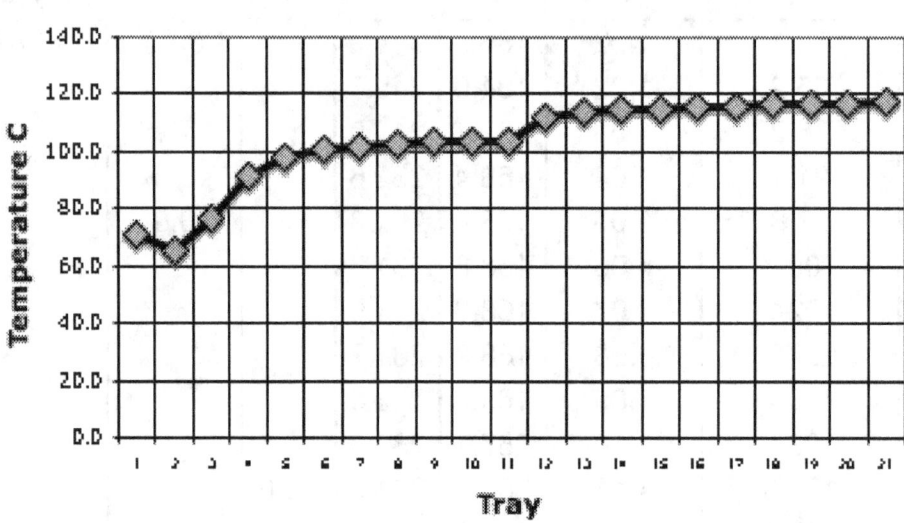

Toluene recovery liquid & vapor flows, tray by tray
Toluene recovery column liquid & vapor flow graph

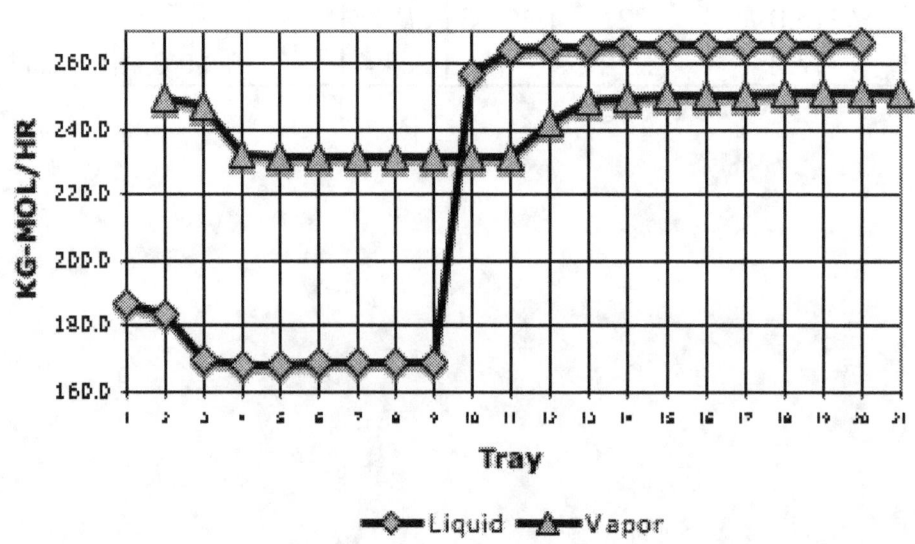

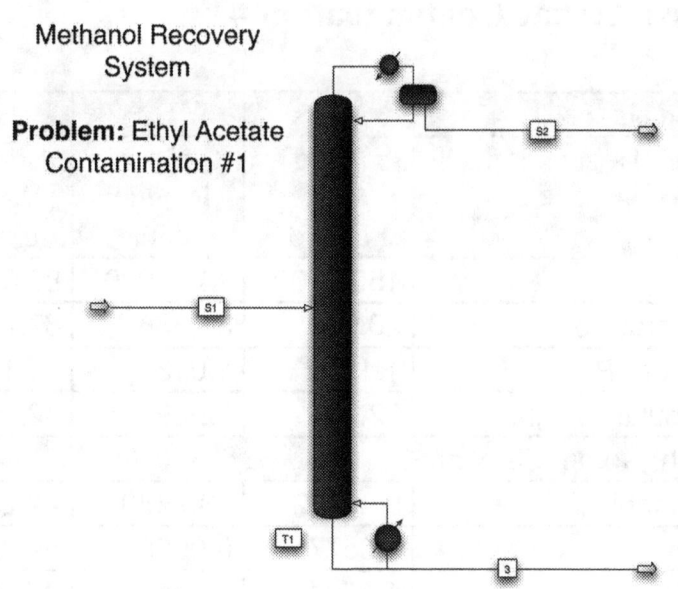

Methanol Recovery
System

Problem: Ethyl Acetate
Contamination #1

X-Y Plot for Methanol and EOAC

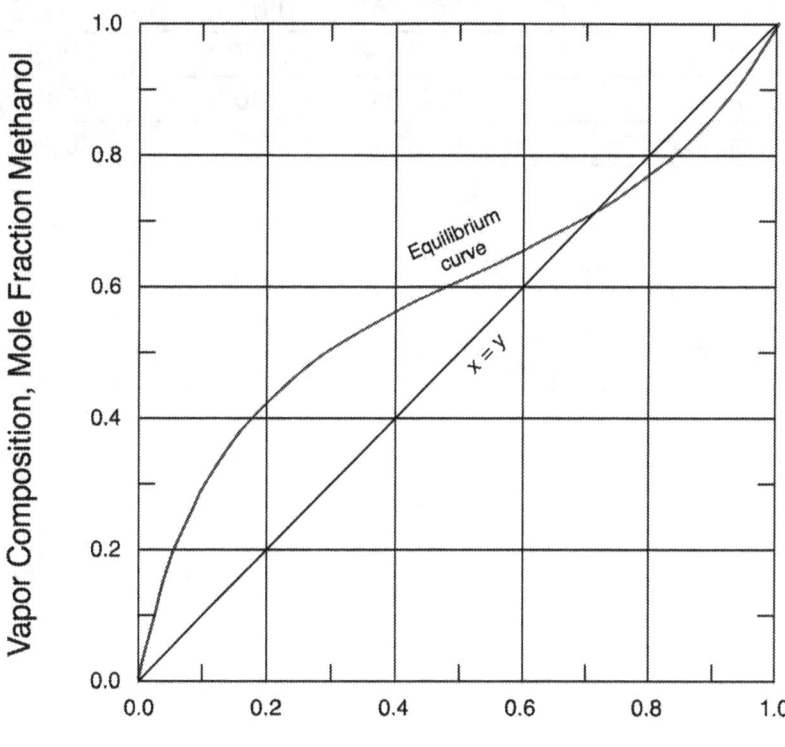

Methanol Recovery System
Problem: Ethyl Acetate Contamination #1

Steam Name	S1	S2	S1
Stream Description Phase	Feed Liquid	Main Product Liquid	Bottom Waste Liquid
KG/HR	1500.000	1415.000	85.000
Temperature C	20.000	64.388	87.705
Pressure BAR	1.100	1.013	1.213
Molecular Weight	32.740	32.299	42.350
Weight Comp. Percent			
Methanol	94.8112	98.7480	29.2660
DMF	3.3577	0.0000	59.2537
EOAC	1.1804	1.2512	0.0004
H2O	0.6507	0.0002	11.4800
Weight Comp. Rates KG/HR			
Methanol	1422.1682	1397.2922	24.8761
DMF	50.3656	0.0000	50.3656
EOAC	17.7054	17.7051	0.0003
H2O	9.7604	0.0026	9.7580
Enthalphy M*KCAL/HR	0.013	0.051	0.006

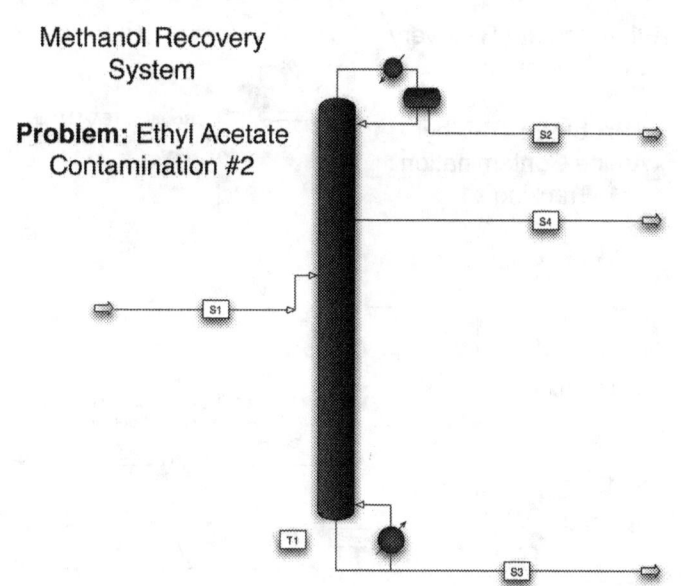

Methanol Recovery
System

Problem: Ethyl Acetate
Contamination #2

Methanol Recovery System
Problem: Ethyl Acetate Contamination #2

Steam Name	S1	S2	S1	S4
Stream Description	Feed	Waste EOAC	Bottom Waste	Main Product
Phase	Liquid	Liquid	Liquid	Liquid
KG/HR	1500.000	200.000	85.000	1215.000
Temperature C	20.000	64.118	87.677	65.216
Pressure BAR	1.100	1.013	1.213	1.045
Molecular Weight	32.740	33.092	42.378	32.171
Weight Comp. Percent				
Methanol	94.8112	95.0130	29.3276	99.359
DMF	3.3577	0.0000	59.2537	0.0000
EOAC	1.1804	4.9869	0.0005	0.6363
H2O	0.6507	0.0001	11.4183	0.0045
Weight Comp. Rates KG/HR				
Methanol	1422.1682	190.0259	24.9284	1207.2139
DMF	50.3656	0.0000	50.3657	0.0000
EOAC	17.7054	9.9737	0.0004	7.7314
H2O	9.7604	0.0003	9.7055	0.0548
Enthalphy M*KCAL/HR	0.013	0.007	0.006	0.044

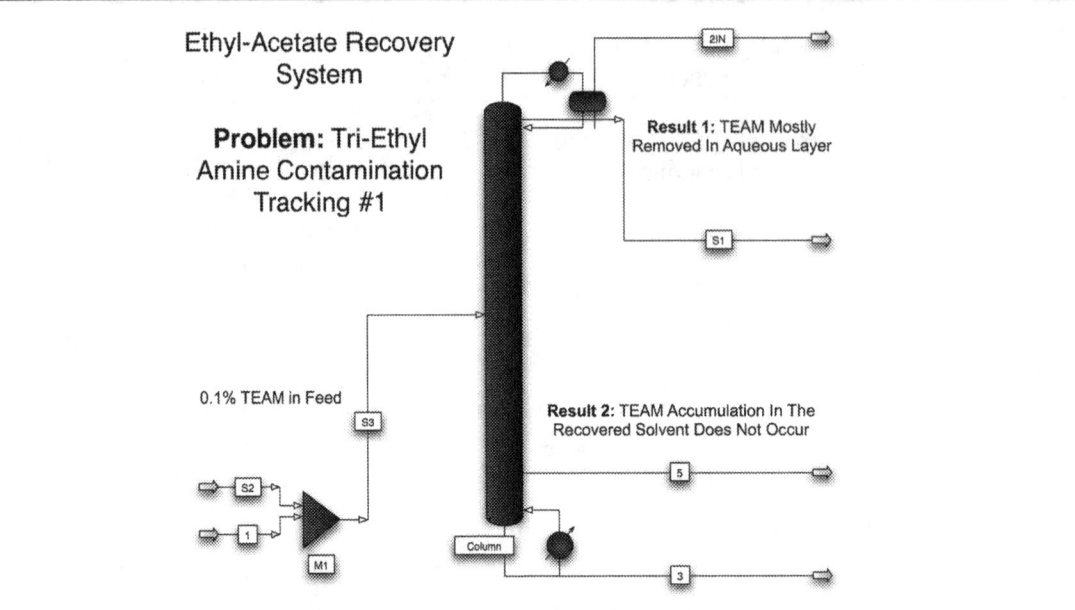

Ethyl-Acetate Recovery System

Problem: Tri-Ethyl Amine Contamination Tracking #1

Result 1: TEAM Mostly Removed In Aqueous Layer

0.1% TEAM in Feed

Result 2: TEAM Accumulation In The Recovered Solvent Does Not Occur

Ethyl Acetate Recovery System
Problem: Tri-Ethyl Amine Contamination Tracking #1

Steam Name	S3	2IN	S1	5	3
Stream Description	Feed	Organic Phase	Water Phase	Main Product	Bottoms Waste
Phase	Liquid	Liquid	Liquid	Vapor	Liquid
KG/HR	3000.000	0.010	127.135	2600.000	272.854
Temperature C	10.016	60.000	60.000	82.141	82.418
Pressure BAR	1.030	1.013	1.013	1.189	1.200
Molecular Weight	76.363	74.837	19.030	88.110	88.108
Weight Comp. Percent					
H2O	3.9560	5.0776	93.3496	0.000	0.0000
EOAC	95.9440	79.2719	5.0150	99.9661	99.9863
TEAM	0.1000	15.6505	1.6349	0.0339	0.0137
DEAM	0.0000	0.0000	0.0000	0.0000	0.0000
Weight Comp. Rates KG/HR					
H2O	118.6812	0.0005	118.6804	0.0003	0.0000
EOAC	2878.3186	0.0079	6.3765	2599.1174	272.8169
TEAM	3.0000	0.0016	2.0786	0.8824	0.0375
DEAM	0.0000	0.0000	0.0000	0.0000	0.0000
Enthalphy M*KCAL/HR	0.014	0.000	0.009	0.326	0.011

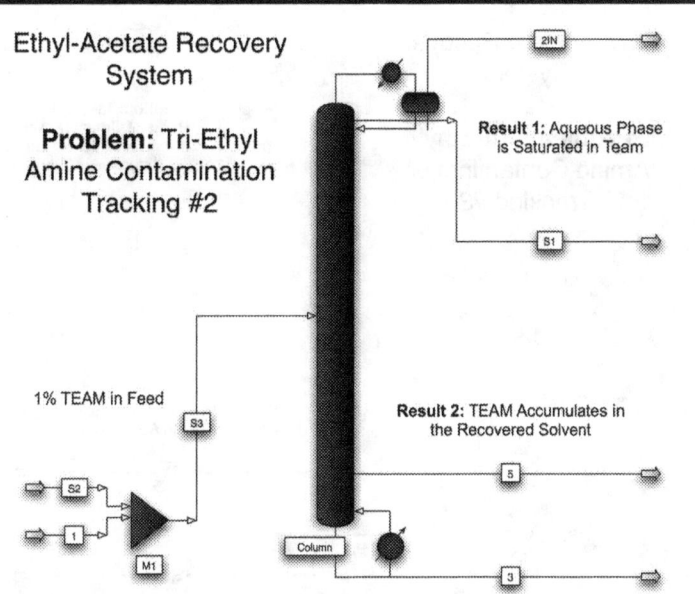

Ethyl Acetate Recovery System
Problem: Tri-Ethyl Amine Contamination Tracking #2

Steam Name	S3	2IN	S1	5	3
Stream Description	Feed	Organic Phase	Water Phase	Main Product	Bottoms Waste
Phase	Liquid	Liquid	Liquid	Vapor	Liquid
KG/HR	3000.000	0.010	126.387	2600.000	273.603
Temperature C	10.163	60.000	60.000	82.018	82.353
Pressure BAR	1.030	1.013	1.013	1.189	1.200
Molecular Weight	76.532	75.963	19.170	88.129	88.148
Weight Comp. Percent					
H2O	3.9204	5.0906	93.5340	0.025	0.0010
EOAC	95.0796	65.3652	4.2816	99.0175	99.6022
TEAM	1.0000	29.5441	3.1844	0.9572	0.3968
DEAM	0.0000	0.0000	0.0000	0.0000	0.0000
Weight Comp. Rates KG/HR					
H2O	117.6120	0.0005	116.9513	0.6576	0.0027
EOAC	2852.3879	0.0065	5.4114	2574.4558	272.5142
TEAM	30.0000	0.0030	4.0246	24.8866	1.0858
DEAM	0.0000	0.0000	0.0000	0.0000	0.0000
Enthalphy M*KCAL/HR	0.014	0.000	0.010	0.327	0.011

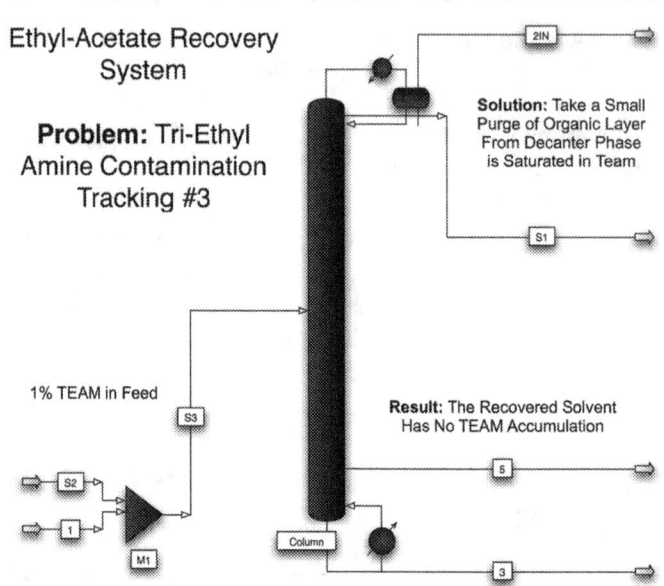

Ethyl-Acetate Recovery System

Problem: Tri-Ethyl Amine Contamination Tracking #3

1% TEAM in Feed

Solution: Take a Small Purge of Organic Layer From Decanter Phase is Saturated in Team

Result: The Recovered Solvent Has No TEAM Accumulation

Ethyl Acetate Recovery System
Problem: Tri-Ethyl Amine Contamination Tracking #3

Steam Name	S3	2IN	S1	5	3
Stream Description	Feed	Organic Phase	Water Phase	Main Product	Bottoms Waste
Phase	Liquid	Liquid	Liquid	Vapor	Liquid
KG/HR	3000.000	80.000	122.328	2600.000	197.672
Temperature C	10.163	60.000	60.000	82.118	82.404
Pressure BAR	1.030	1.013	1.013	1.189	1.200
Molecular Weight	76.532	75.575	19.170	88.136	89.118
Weight Comp. Percent					
H2O	3.9204	5.0856	92.8185	0.000	0.0000
EOAC	95.0796	70.1248	4.5384	99.7412	99.8949
TEAM	1.0000	24.7896	2.6431	0.2587	0.1051
DEAM	0.0000	0.0000	0.0000	0.0000	0.0000
Weight Comp. Rates KG/HR					
H2O	117.6120	4.0685	113.5433	0.0002	0.0000
EOAC	2852.3879	56.0998	5.5518	2593.2722	197.4639
TEAM	30.0000	19.8317	3.2333	6.7274	0.2077
DEAM	0.0000	0.0000	0.0000	0.0000	0.0000
Enthalphy M*KCAL/HR	0.014	0.003	0.009	0.326	0.008

Consider how to design & rate another multi-purpose distillation system, this time required to process at least nine different feed materials

- Weekly production rates govern overall system flow capacity
- Product can be light or heavy phase from decanter, or liquid side draw, or vapor side draw
- Batch operation must be possible, along with continuous operation
- Product residue of evaporation must be very low
- Maximum fractionation (removing methylene chloride from acetone) requires 30 theoretical trays, 60 real valve trays will be provided
- Column will be split to reduce overall height for better wind rating & possible simultaneous dual column operation

Project 98J03
Solvent Recovery Typical Annual Operational Requirements

System #1 Process Code C6

Recoverable Solvent:	Typical Feed Contaminants Weight %:
Acetonitrile (ACN)	**32% MTBE**
Recovery Quality Specs (Prelim):	Annual Recovery Amount US Gal:
<0.5% MTBE, 99% ACN	**152,471**
Peak Weekly Amount US Gal:	Typical Feed Rate US Gal/Min:
4,646	**14.38**
Recovery Hours On-Stream:	
175/Year, 5.4/Week	

System #2 Process Code C3

Recoverable Solvent:	Typical Feed Contaminants Weight %:
Methyl Tert-Butyl Ether (MTBE)	**22% MDC**
Recovery Quality Specs (Prelim):	Annual Recovery Amount US Gal:
<0.1% MDC, 99.9% MTBE	**78,553**
Peak Weekly Amount US Gal:	Typical Feed Rate US Gal/Min:
5,395	**6.46**
Recovery Hours On-Stream:	
203/Year, 13.9/Week	

System #3 Process Code S1

Recoverable Solvent:	Typical Feed Contaminants Weight %:
Isopropanol (IPA)	**2.96% Water, 0.11% Methanol, 0.46 % Acetone**
Recovery Quality Specs (Prelim):	Annual Recovery Amount US Gal:
<0.4% Water, <0.01% Methanol, <0.01% Acetone	**327,116**
Peak Weekly Amount US Gal:	Typical Feed Rate US Gal/Min:
14,076	**8.15**
Recovery Hours On-Stream:	
709/Year, 29/Week	

System #4 Process Code C2

Recoverable Solvent:	Typical Feed Contaminants Weight %:
Acetone	**11.6% IPA**
Recovery Quality Specs (Prelim):	Annual Recovery Amount US Gal:
<0.1% IPA, 99.0% Acetone	**313,359**
Peak Weekly Amount US Gal:	Typical Feed Rate US Gal/Min:
13,524	**6.29**
Recovery Hours On-Stream:	
830/Year, 36/Week	

System #5 Process Code S2

Recoverable Solvent:	Typical Feed Contaminants Weight %:
Acetone	**0.21% Methanol, 0.30% IPA, 0.82% MDC, 1.30% Water**
Recovery Quality Specs (Prelim):	Annual Recovery Amount US Gal:
0.-3% Methanol, 0.10% IPA, 0.20% MDC, 0.67% Water, 99.0% Acetone	**313,359**
Peak Weekly Amount US Gal:	Typical Feed Rate US Gal/Min:
13,524	**4.58**
Recovery Hours On-Stream:	
2,280/Year, 98/Week	

System #6 Process Code S3

Recoverable Solvent:	Typical Feed Contaminants Weight %:
Toluene	**0.48% Methanol, 0.05% IPA, 0.01% MDC, 0.51% Water**
Recovery Quality Specs (Prelim):	Annual Recovery Amount US Gal:
<0.01% Methanol, <0.01% IPA, <0.01% MDC, <0.05% Water, 99.9% Acetone	**123,049**
Peak Weekly Amount US Gal:	Typical Feed Rate US Gal/Min:
7,537	**10.71**
Recovery Hours On-Stream:	
191/Year, 12/Week	

System #7 Process Code S4

Recoverable Solvent:	Typical Feed Contaminants Weight %:
Methylene Dichloride (MDC)	**0.11% Methanol, 0.31% Water**
Recovery Quality Specs (Prelim):	Annual Recovery Amount US Gal:
<0.01% Methanol, 0.05% Water	**239,165**
Peak Weekly Amount US Gal:	Typical Feed Rate US Gal/Min:
15,132	**5.96**
Recovery Hours On-Stream:	
669/Year, 42/Week	

System #8

Recoverable Solvent:	Recovery Quality Specs (Prelim):
Ethyl Acetate (ETAC)	**Undefined**
Typical Feed Contaminants:	
Undefined Organics Plus Water	
Annual Recovery Amount US Gal:	Peak Weekly Amount US Gal:
78,439	**4,804**

System #9

Recoverable Solvent:	Recovery Quality Specs (Prelim):
Methyl Isobutyl Ketone (MIBK)	**Undefined**
Typical Feed Contaminants:	
Undefined Organics Plus Water	
Annual Recovery Amount US Gal:	Peak Weekly Amount US Gal:
24,218	**14,676**

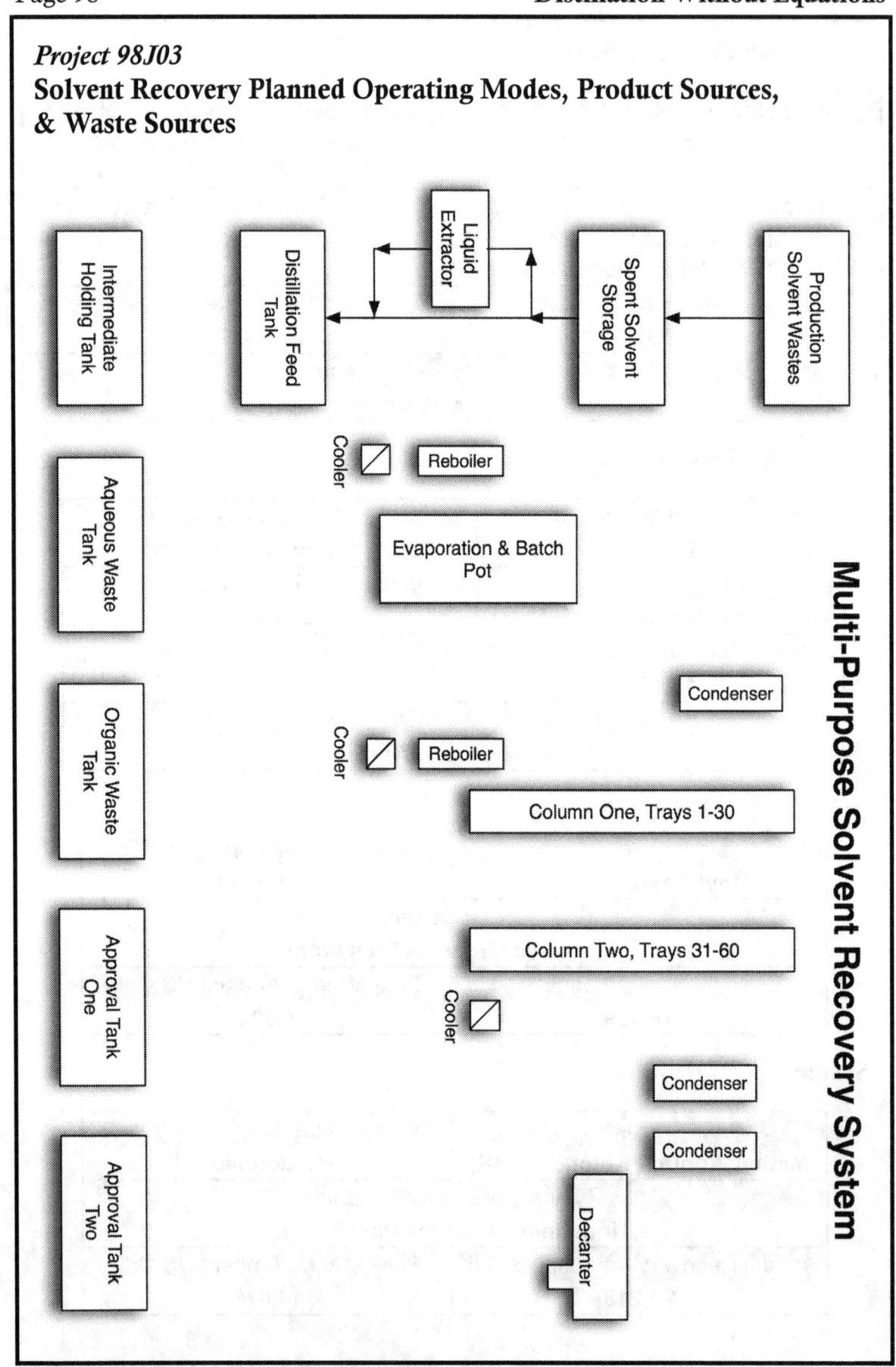

Project 98J03
Solvent Recovery Planned Operating Modes, Product Sources,
& Waste Sources

System #1 Process Code C6

Recoverable Solvent:	Operation Type:
ACN	**Continuous Using Batch Pot**
Decanter Mode:	Product Source:
Light	**Batch Pot to VSD Condenser**
Waste Sources:	
Batch Pot Bottoms, Decanter Light and Heavy	

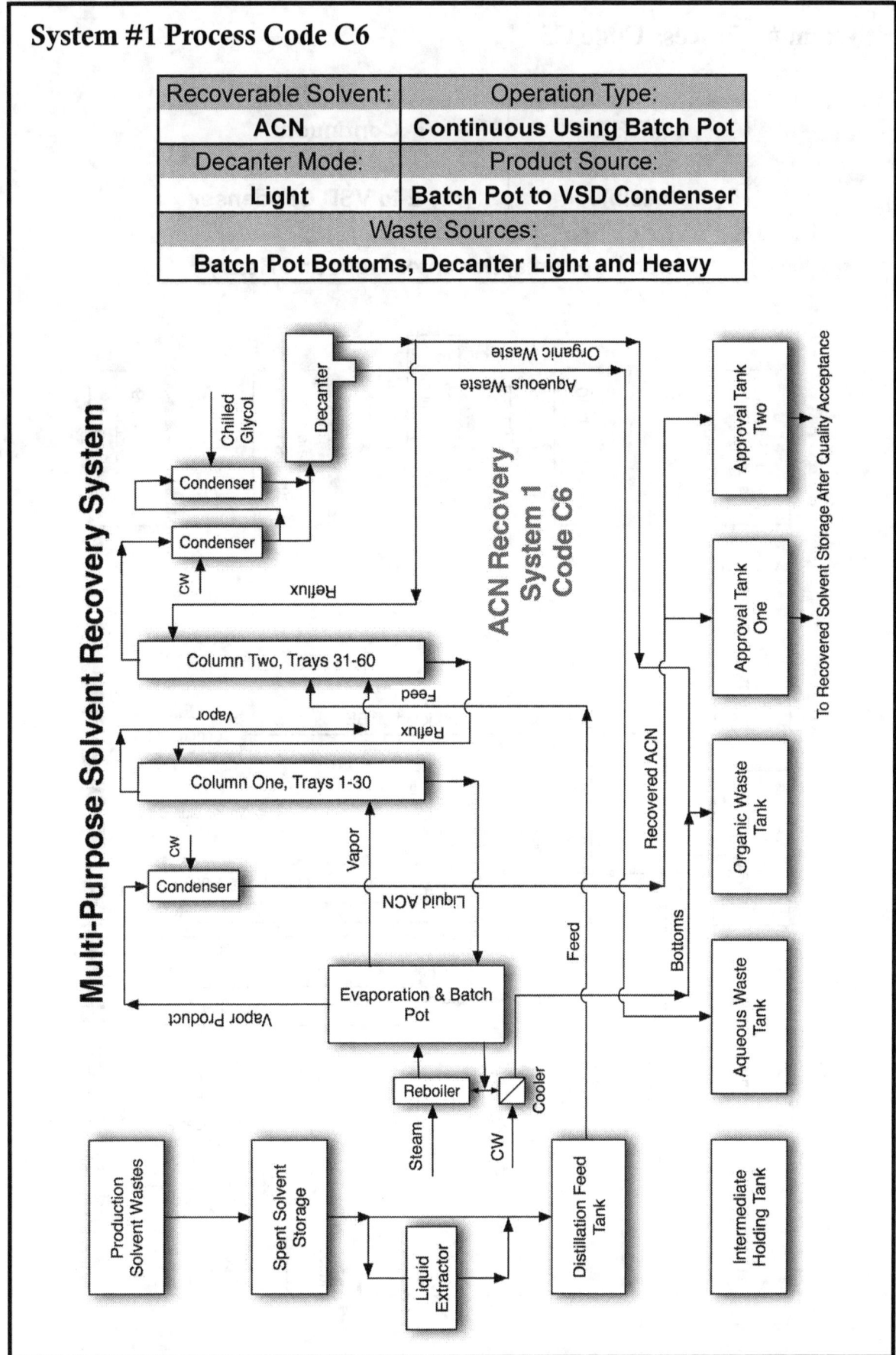

System #2 Process Code C3

Recoverable Solvent:	Operation Type:
MTBE	**Continuous**
Decanter Mode:	Product Source:
Misible	**Tray 2 to VSD Condenser**
Waste Sources:	
Tower 1 Bottoms, Decanter Light and Heavy	

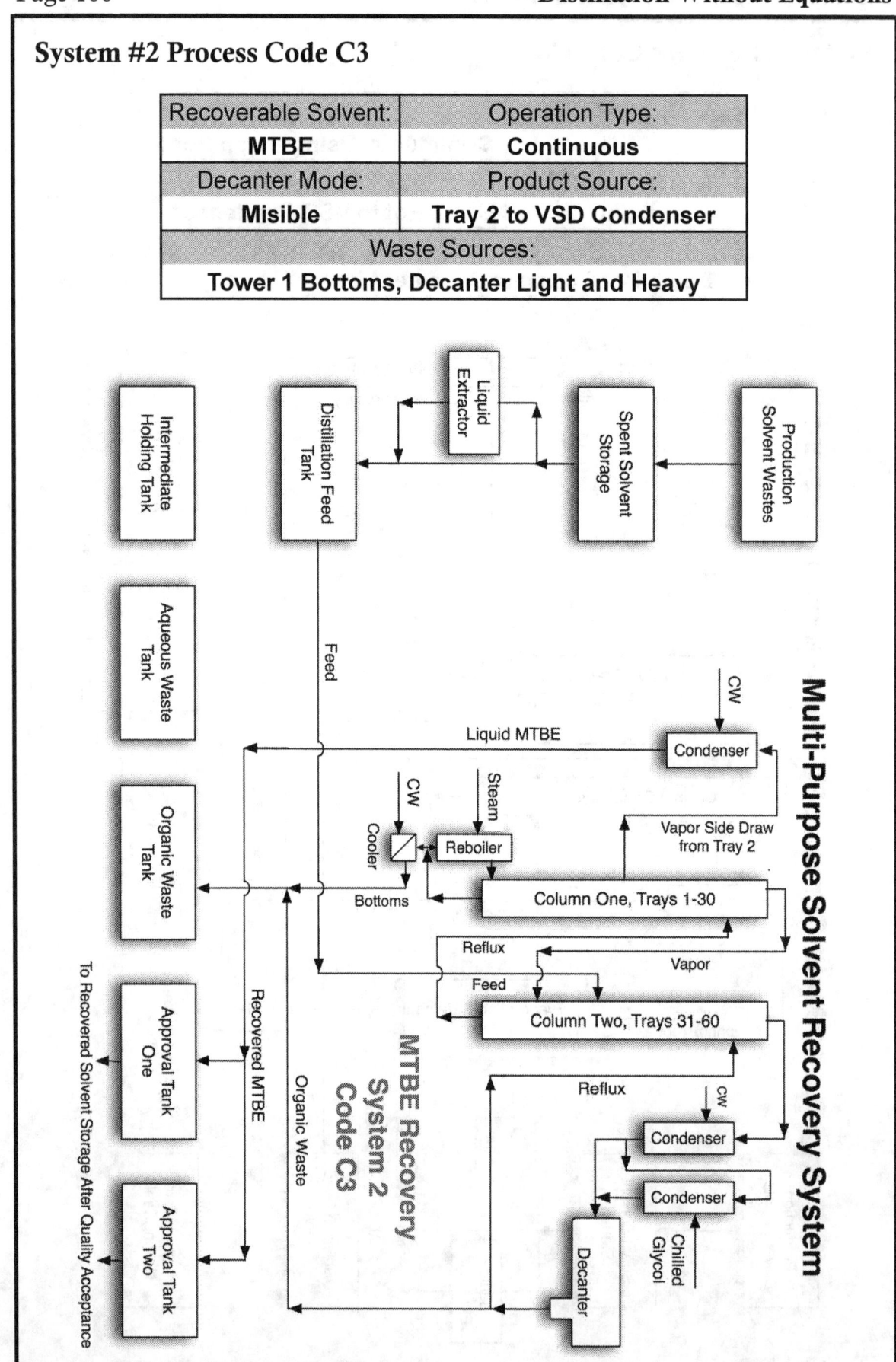

System #3 Process Code S1

Recoverable Solvent:	Operation Type:
IPA	**Continuous Using Batch Pot**
Decanter Mode:	Product Source:
Misible	**Tray 2 to VSD Condenser**
Waste Sources:	
Batch Pot Bottoms, Decanter Heavy	

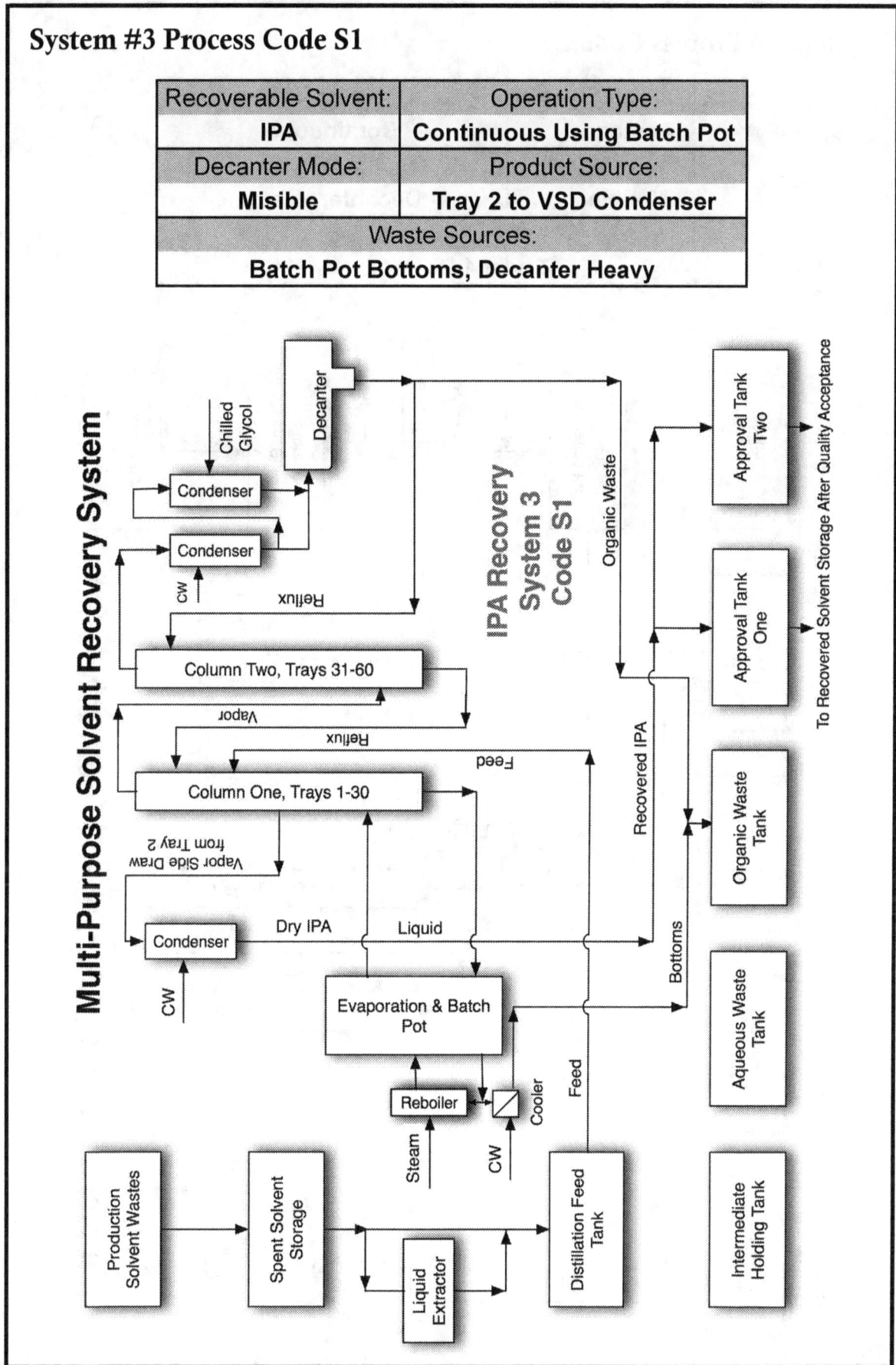

System #4 Process Code C2

Recoverable Solvent:	Operation Type:
Acetone	**Continuous**
Decanter Mode:	Product Source:
Misible	**Decanter Heavy**
Waste Sources:	
Tower 1 Bottoms	

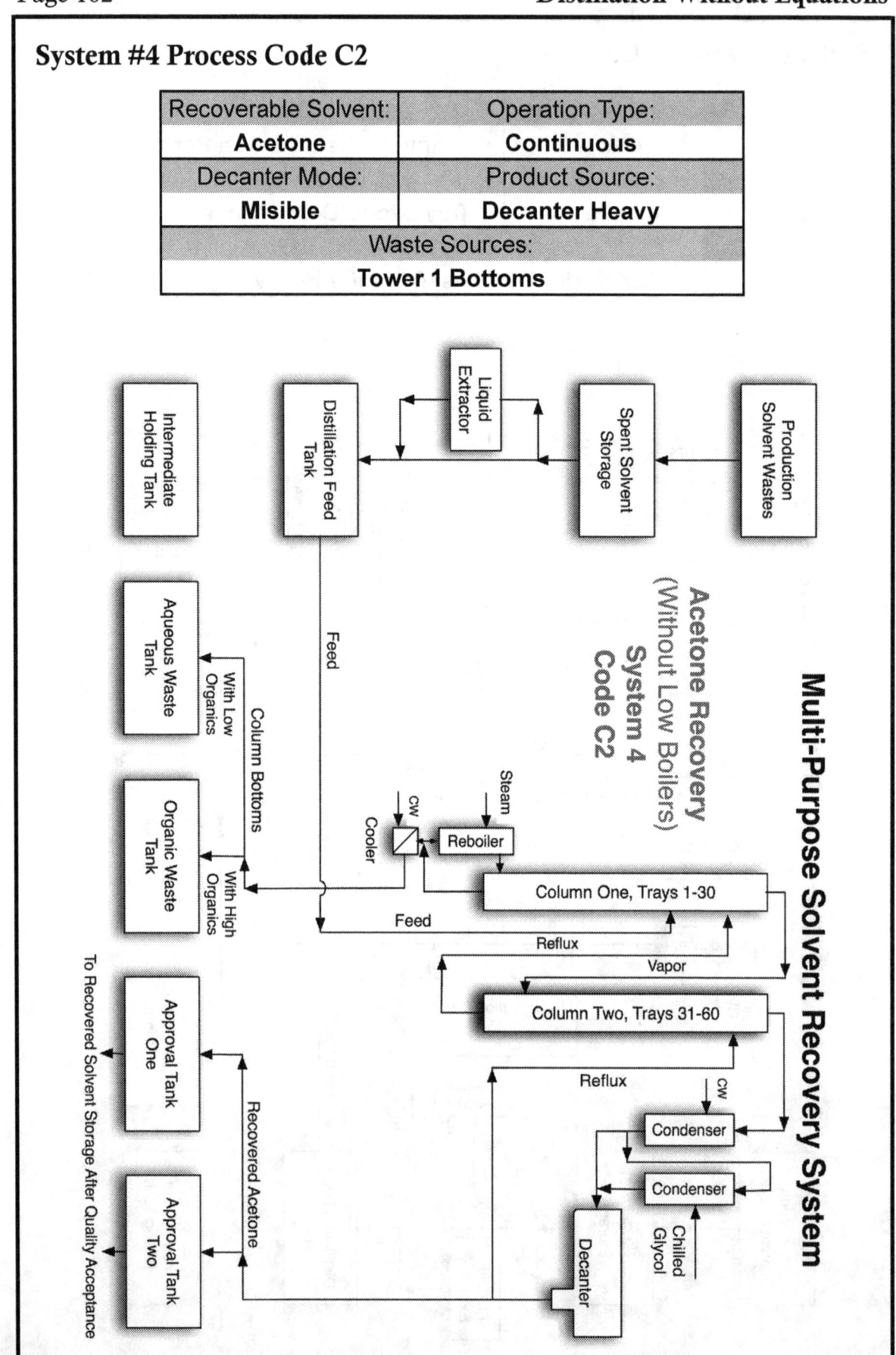

System #5 Process Code S2

Recoverable Solvent:		Operation Type:
Acetone	1st Pass	**Continuous**
Decanter Mode:		Product Source:
Misible		**Decanter Heavy**
Waste Sources:		
Tower 1 Bottoms		

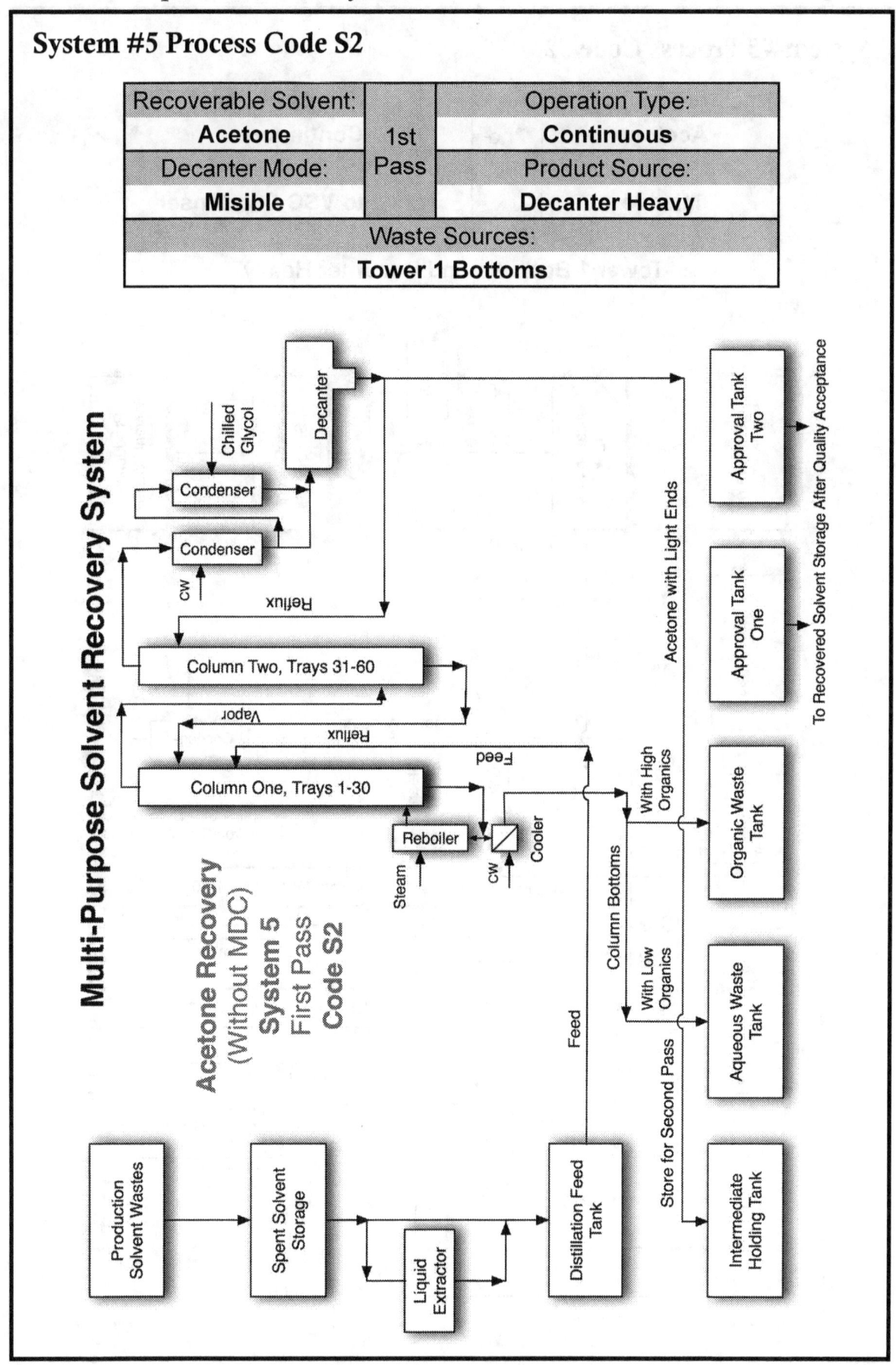

Multi-Purpose Solvent Recovery System

Acetone Recovery
(Without MDC)
System 5
First Pass
Code S2

System #5 Process Code S2

Recoverable Solvent:	2nd Pass	Operation Type:	
Acetone		**Continuous**	
Decanter Mode:		Product Source:	
Misible		**Tray 2 to VSC Condenser**	
Waste Sources:			
Tower 1 Bottoms and Decanter Heavy			

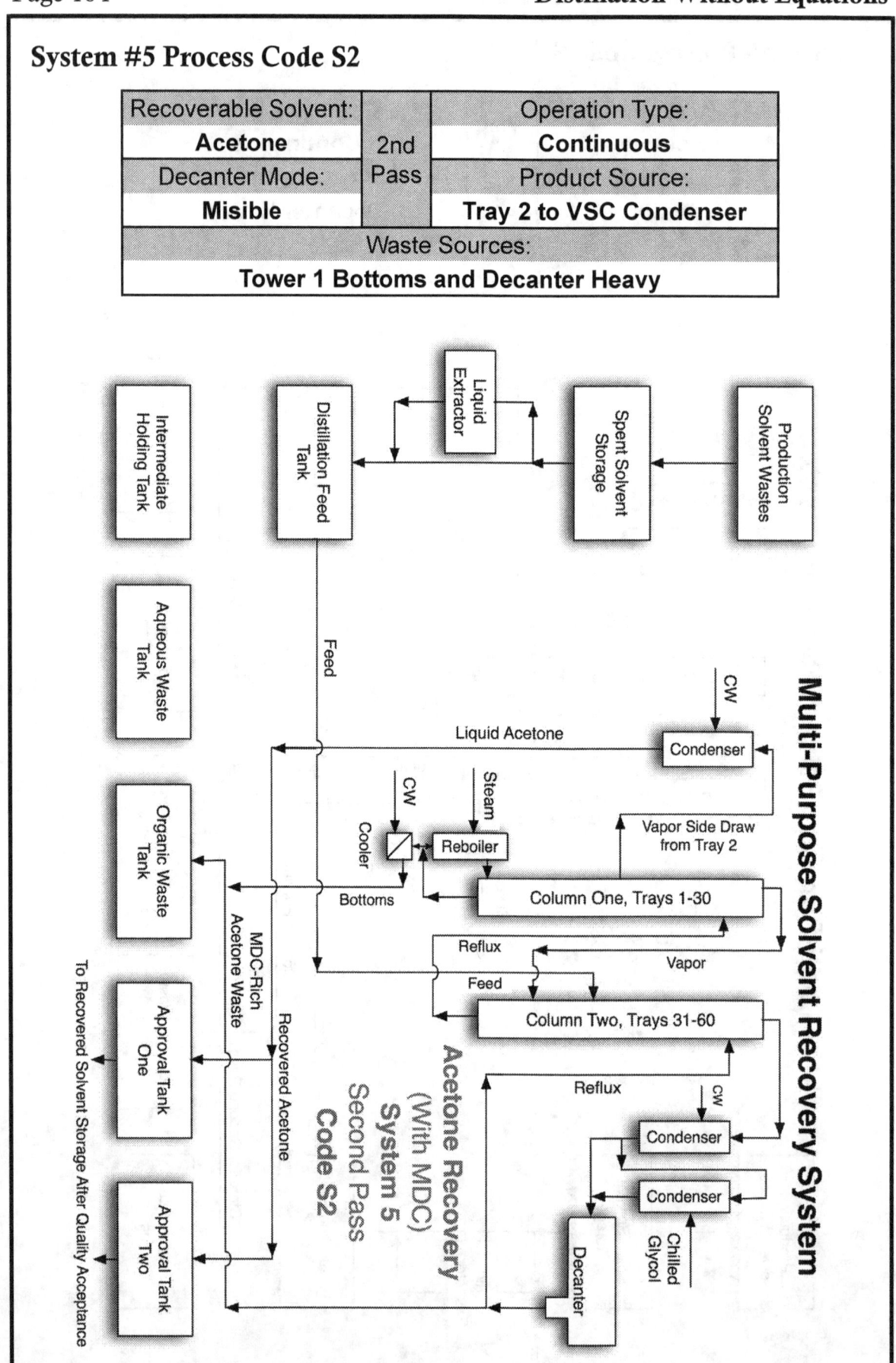

System #6 Process Code S3

Recoverable Solvent:	Operation Type:
Toluene	**Continuous Using Batch Pot**
Decanter Mode:	Product Source:
Light	**Batch Pot to VSD Condenser**
Waste Sources:	
Tower 1 Bottoms, Decanter Light and Heavy	

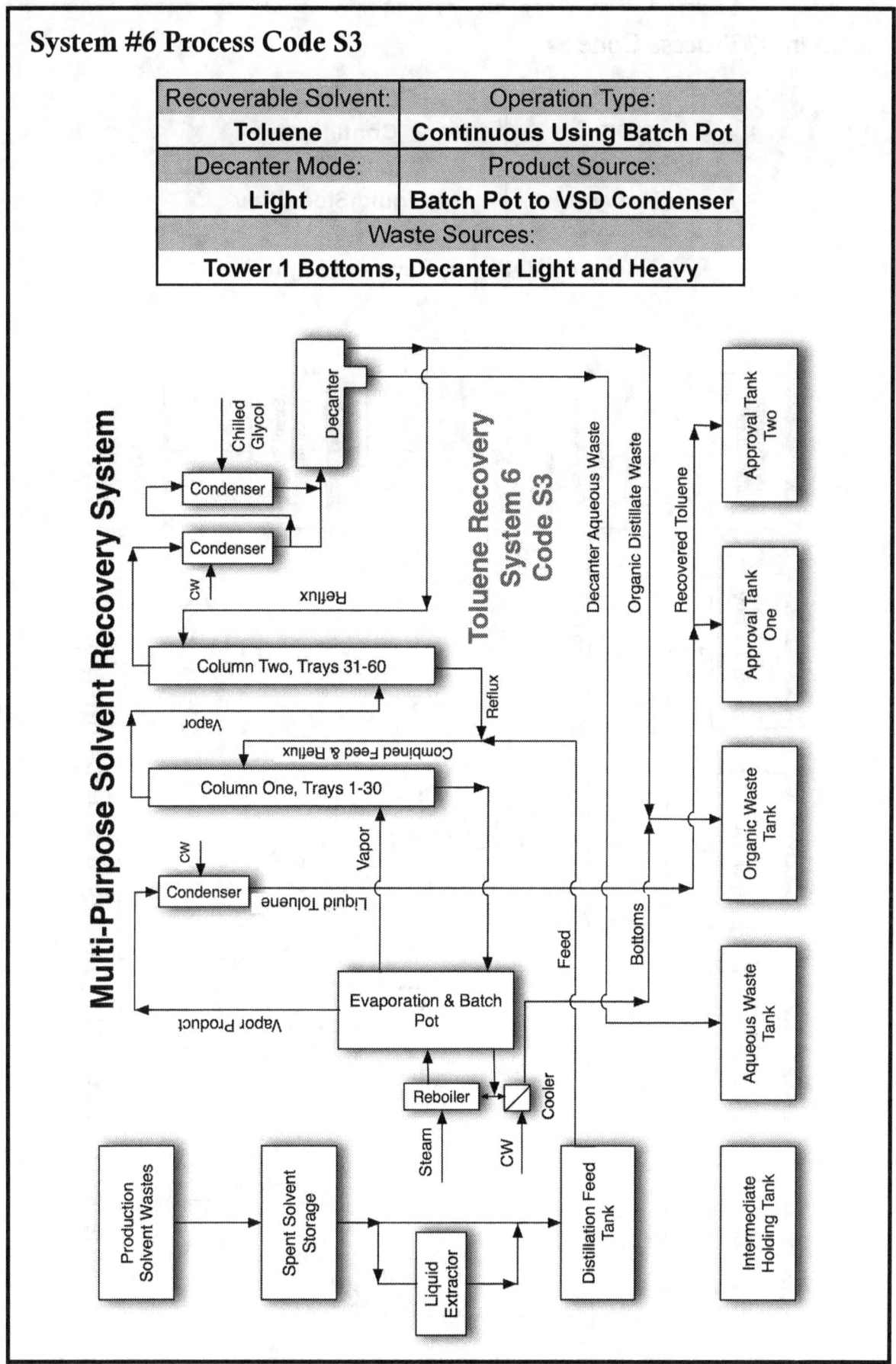

System #7 Process Code S4

Recoverable Solvent:	Operation Type:
MDC	**Continuous**
Decanter Mode:	Product Source:
Heavy	**Liquid Side Draw**
Waste Sources:	
Tower 1 Bottoms, Decanter Light	

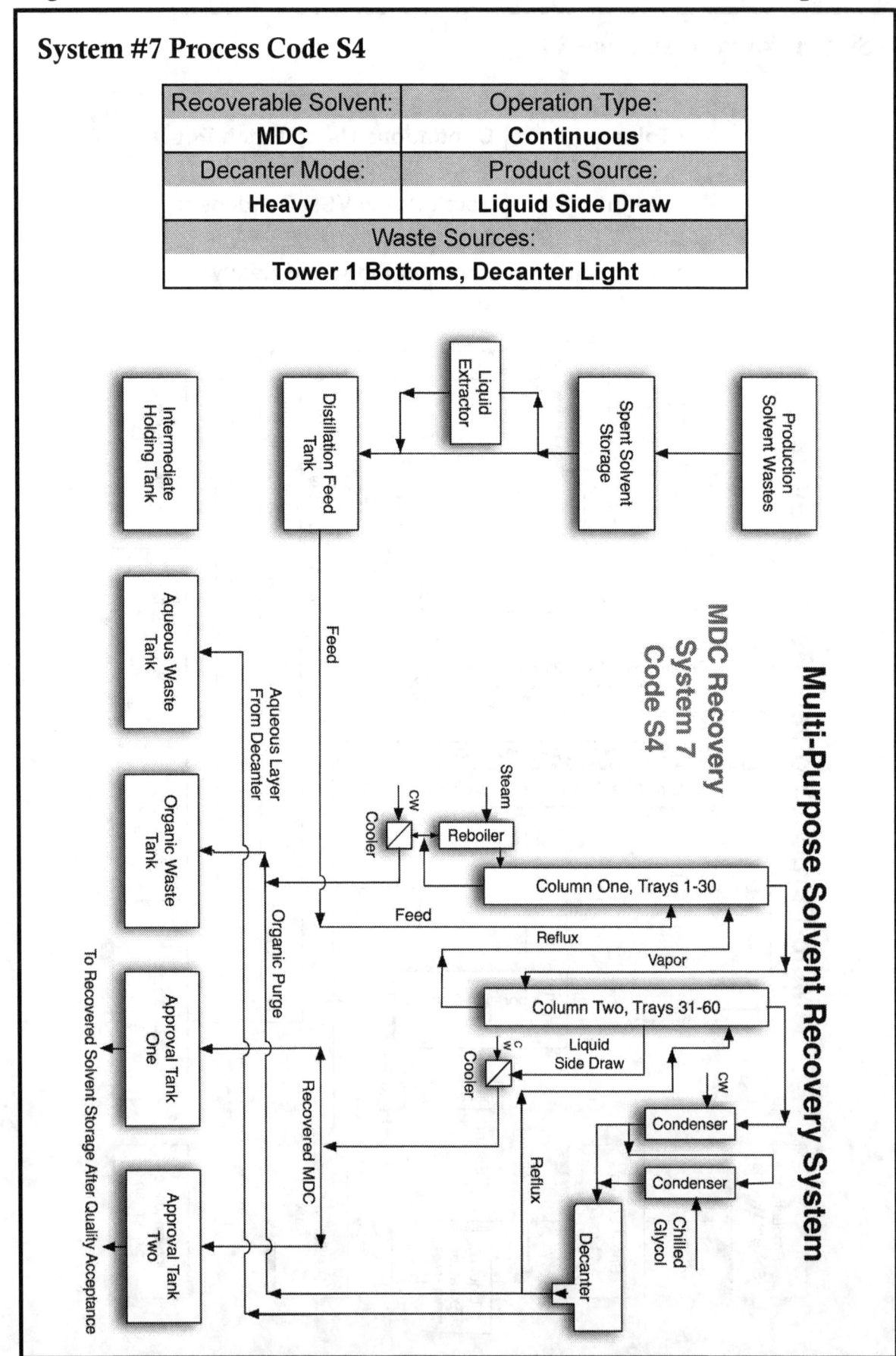

System #8

Recoverable Solvent:	Operation Type:
ETAC	**Continuous**
Decanter Mode:	Product Source:
Light	**Tray 2 to VSD Condenser**
Waste Sources:	
Decanter Light and Heavy, Liquid Side Draw, Tower 1 Bottoms	

System #9

Recoverable Solvent:	Operation Type:
MIBK	**Continuous**
Decanter Mode:	Product Source:
Light	**Liquid Side Draw**
Waste Sources:	
Tower 1 Bottoms, Decanter Light and Heavy	

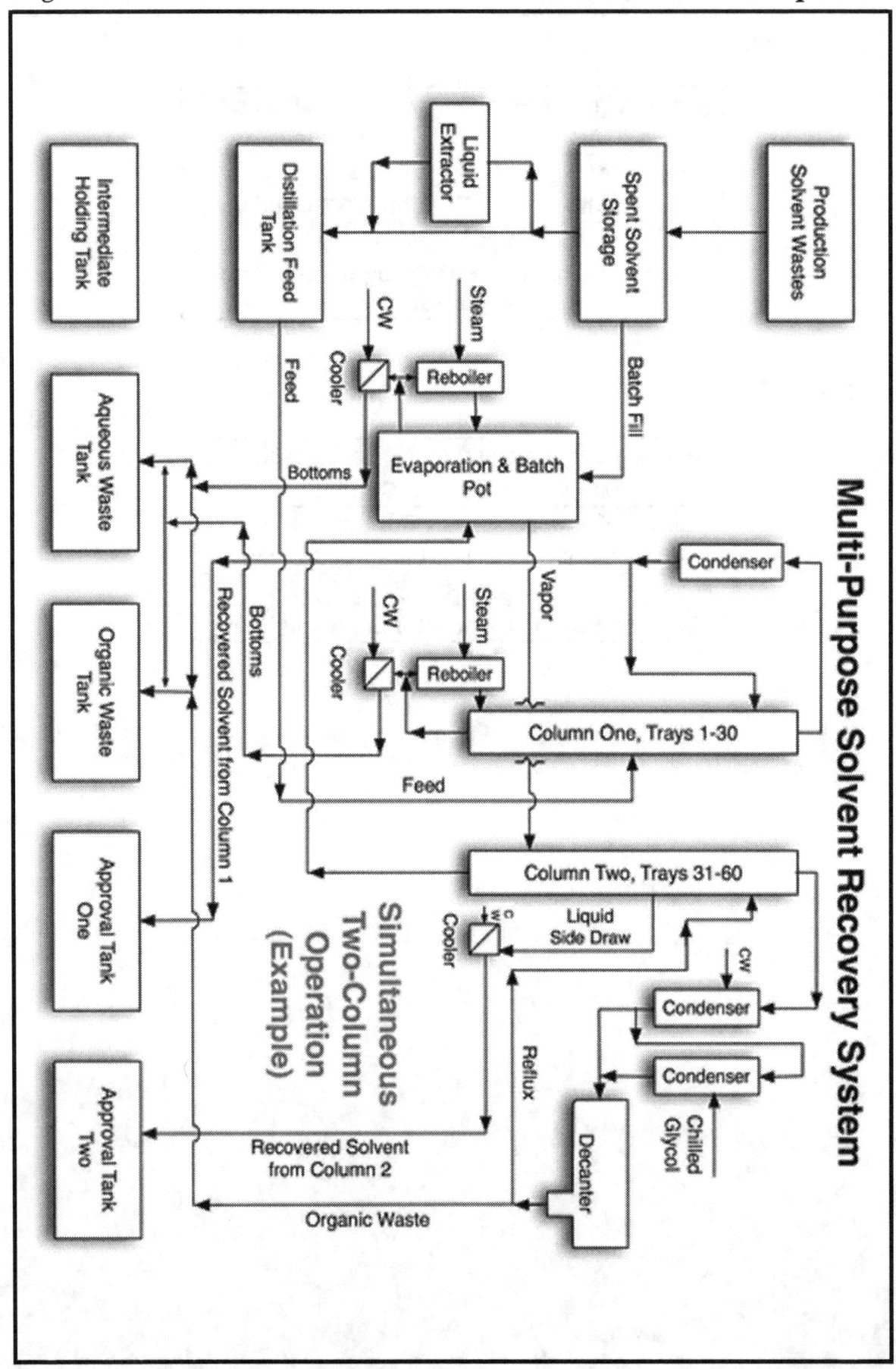

Lab simulation of a commercial scale distillation system

Laboratory distillation apparatus mimics commercial systems

To fit in the lab, one theoretical tray per 1 to 4 inches of column height is commonly required

Lab requires tiny distillation trays or tiny packing

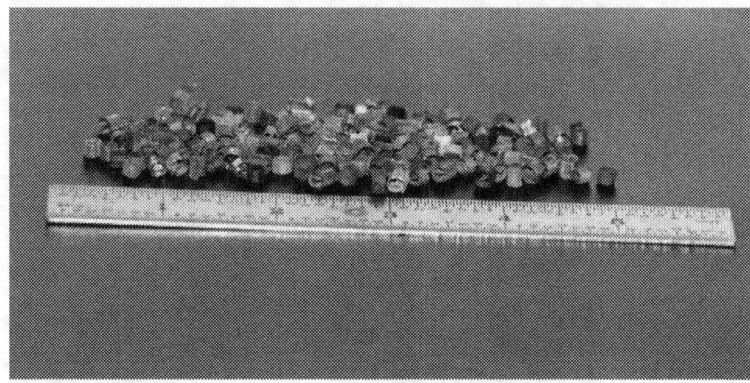

Distillation lab analytical requirements include GC, HPLC, moisture balance, Karl Fischer

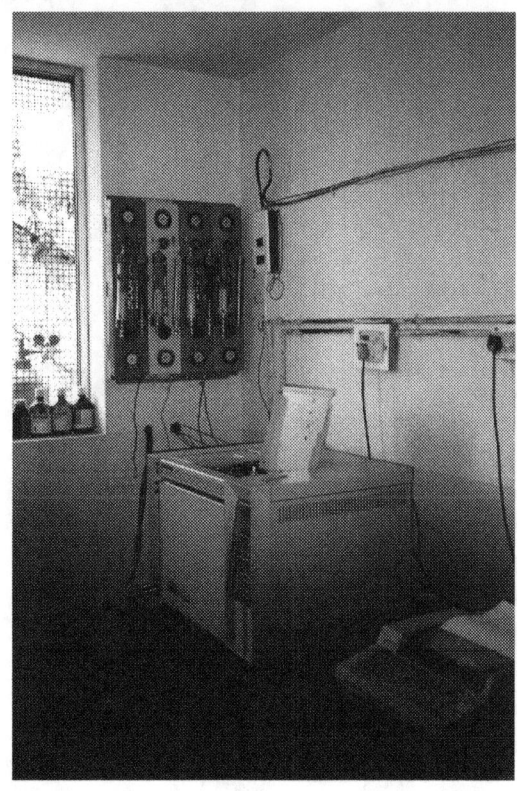

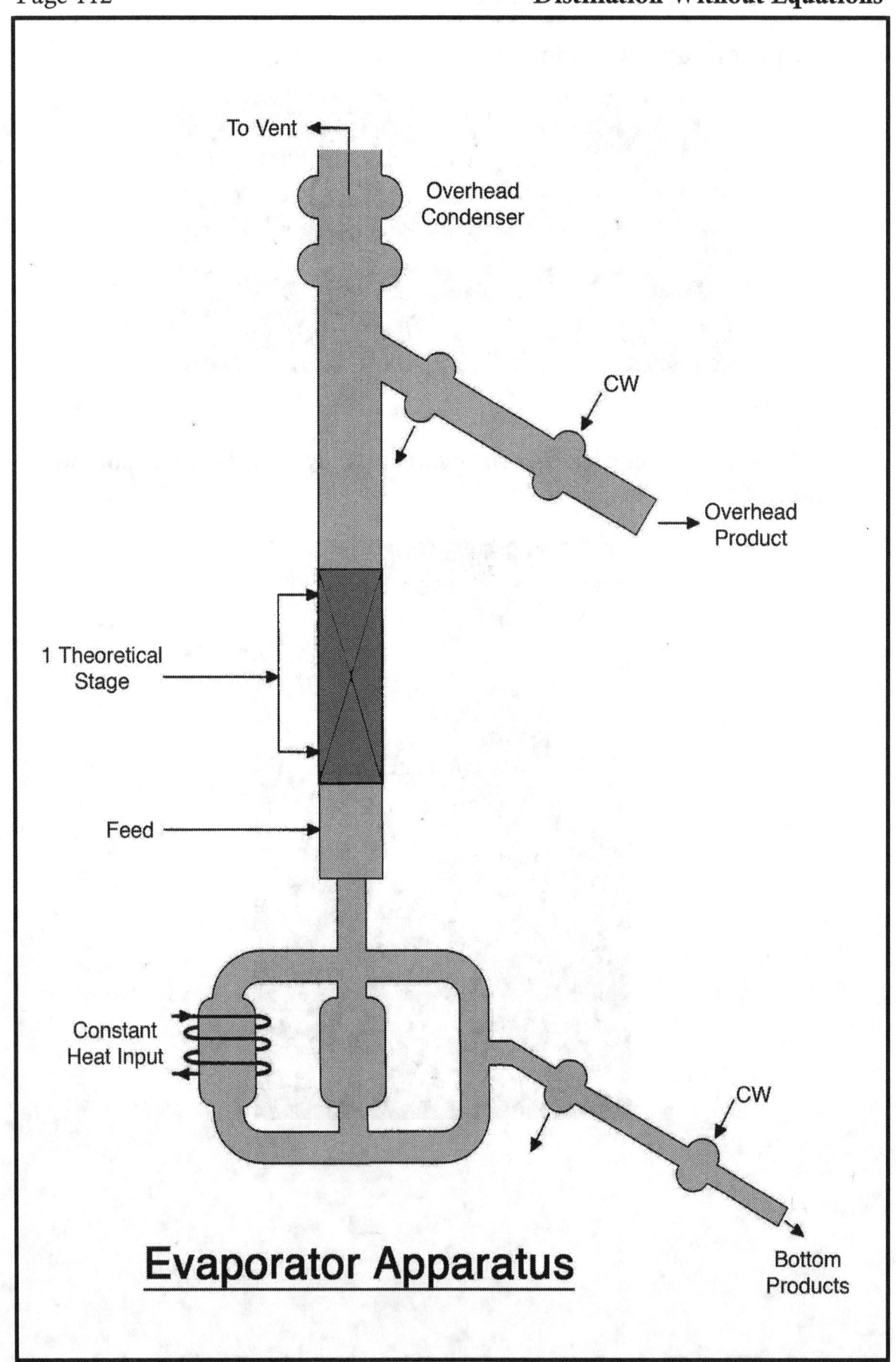

Evaporator Apparatus

Computer simulation flowsheet model of of IPA evaporator

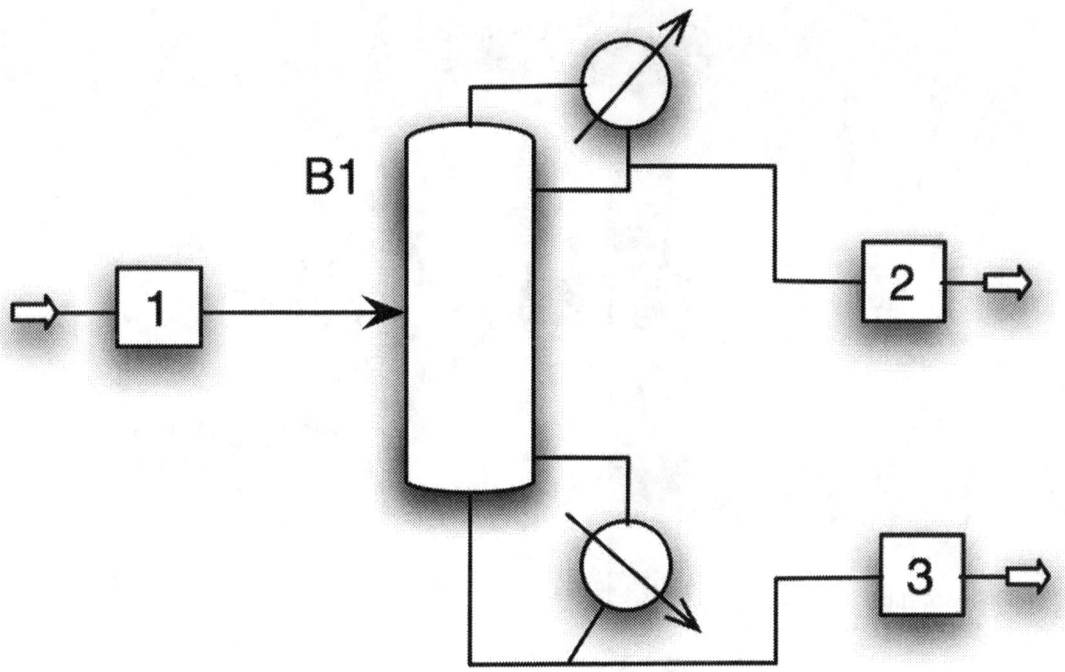

Lab IPA evaporator computer simulation stream data

IPA/DBP/Water			
Stream ID	1	2	3
Temperature C	15.0	82.2	83.7
Mass Flow KG/HR	800.000	710.000	90.000
Mass Fraction			
Water	0.002	0.002	81 PPB
ISOPR-01	0.983	0.998	0.867
METHA-01			
DIBUT-01	0.015	trace	0.133

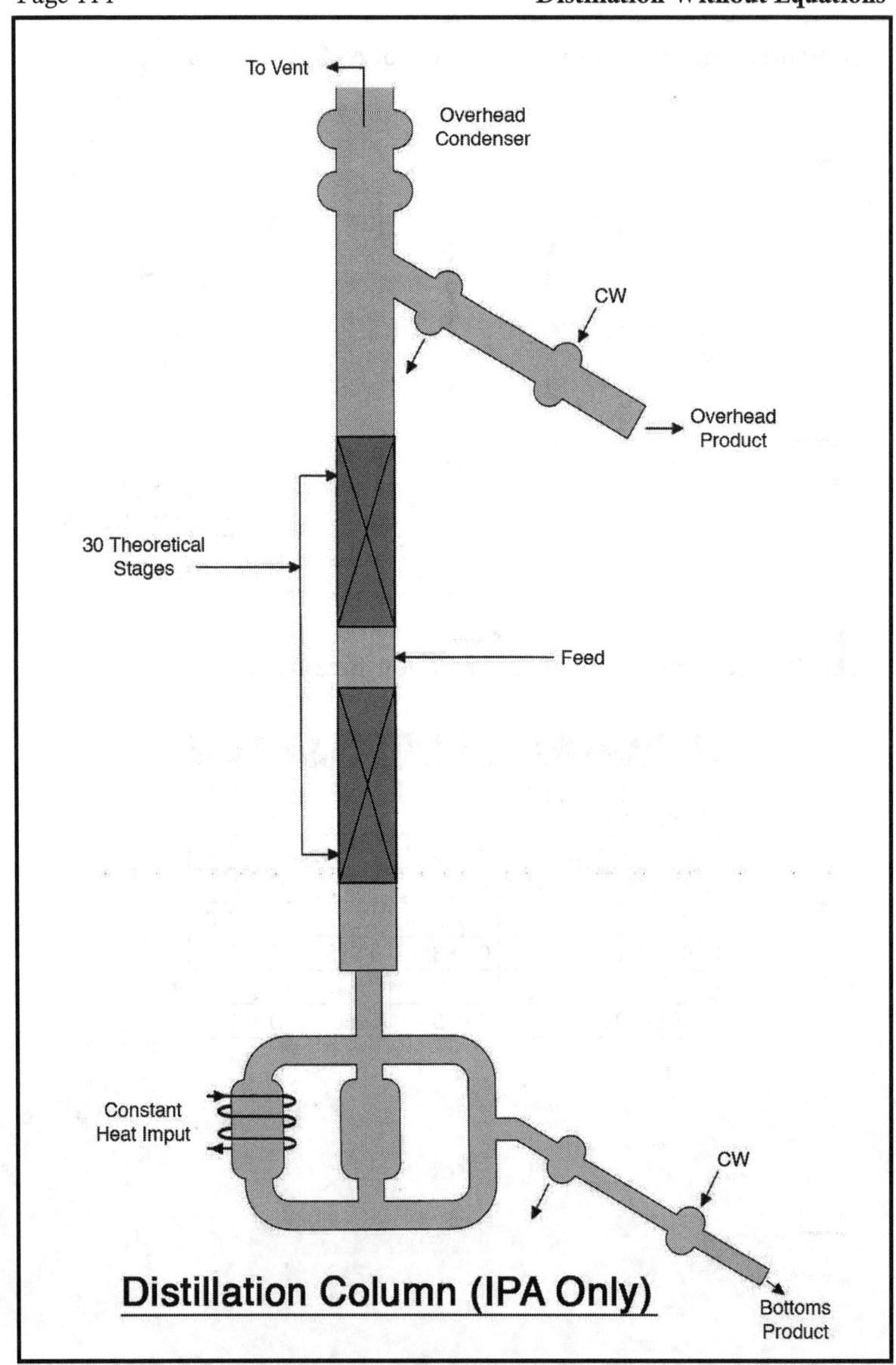

Distillation Column (IPA Only)

Lab IPA - water distillation computer simulation model

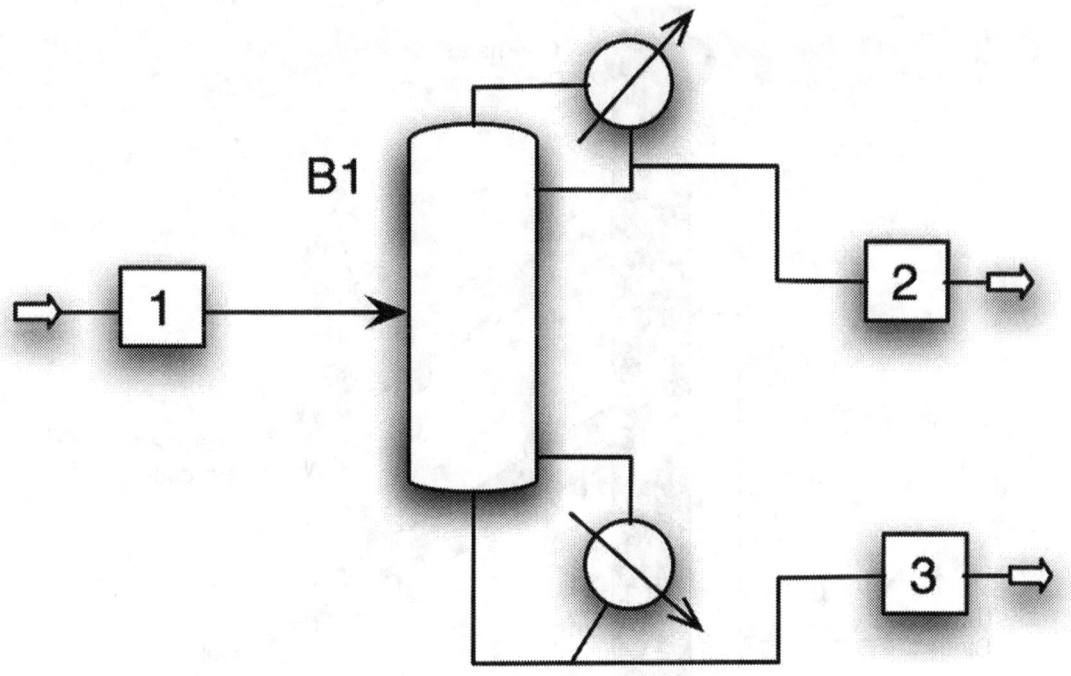

Lab IPA distillation computer simulation stream data

IPA/Water			
Stream ID	1	2	3
Temperature (C)	15.0	81.9	82.8
Mass Flow (KG/HR)	800.000	150.924	649.079
Mass Fraction			
Water	0.002	0.010	100 PPM
ISOPR-01	0.998	0.990	1.000
METHA-01			
DIBUT-01			

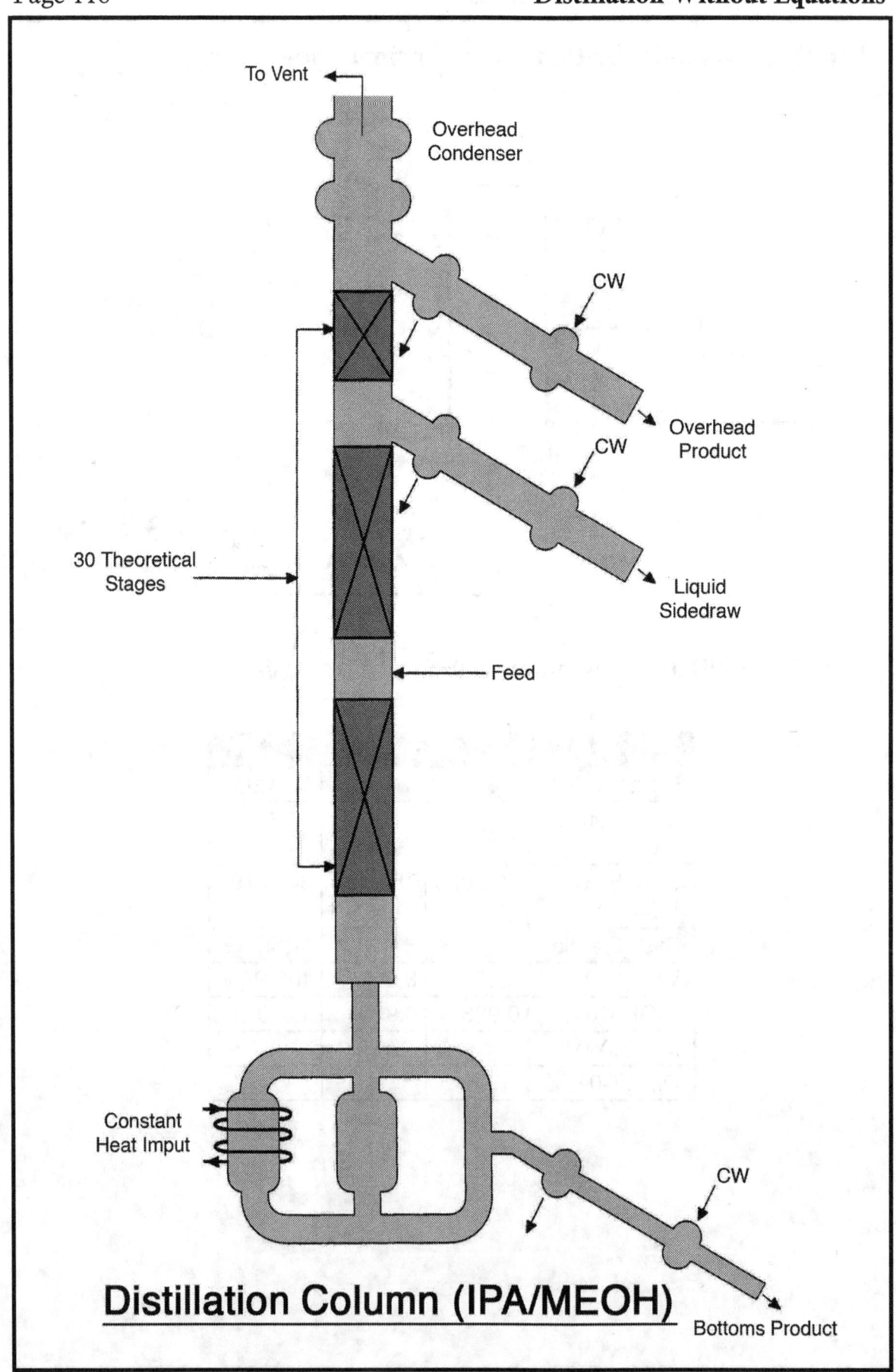

Distillation Column (IPA/MEOH)

Lab IPA - methanol - water computer simulation model

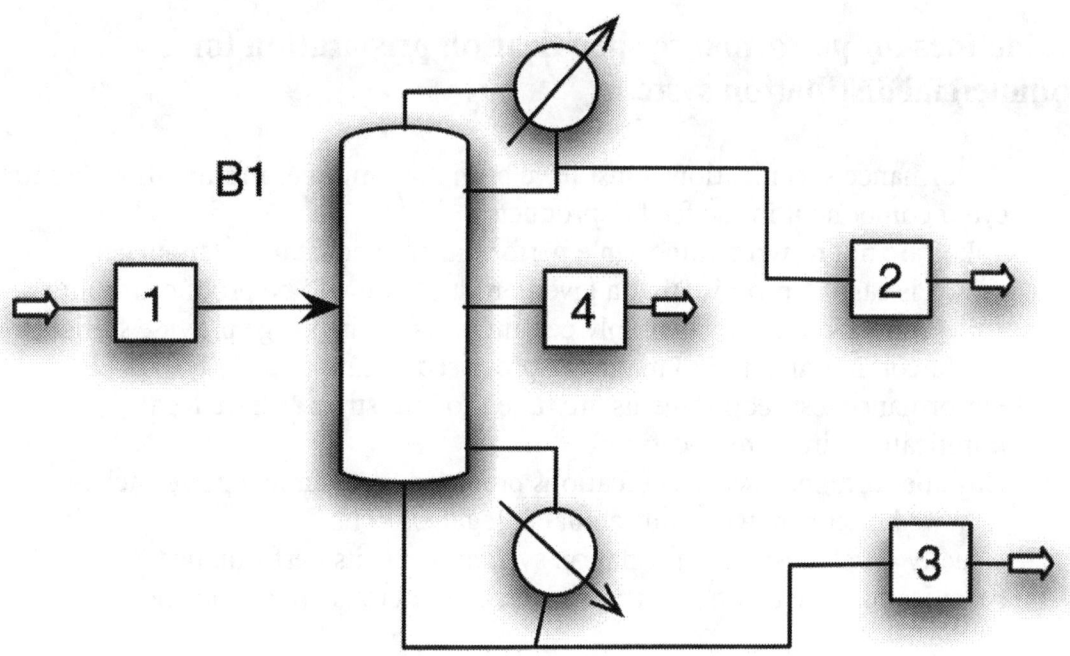

Lab IPA - methanol - water computer simulation stream data

IPA/Methanol/Water				
Stream ID	1	2	3	4
Temperature (C)	15.0	66.3	82.8	74.6
Mass Flow (KG/HR)	500.000	27.000	446.000	27.000
Mass Fraction				
Water	0.002	0.008	102 PPM	0.028
ISOPR-01	0.948	0.249	1.000	0.790
METHA-01	0.050	0.744	918 PPB	0.182
DIBUT-01				

Performance testing of a commercial distillation system

Guidelines for performance specification preparation for a commercial distillation system

- Performance specifications must have both maximum & minimum values for every component in the feed & products
- Take care not to write impossible performance specification language:
    - It is easy to promise that a given product rate will be produced from a given feed range, when only certain parts of the range provide sufficient incoming material to make the promised output
- Performance test requirements are tuned to industries & have legal ramifications in many locations
- Have performance test specifications prepared by technical personnel but reviewed by others with contractual & legal expertise
- Usually a well designed distillation system meets its performance specifications easily on first try, if the feed materials are as specified

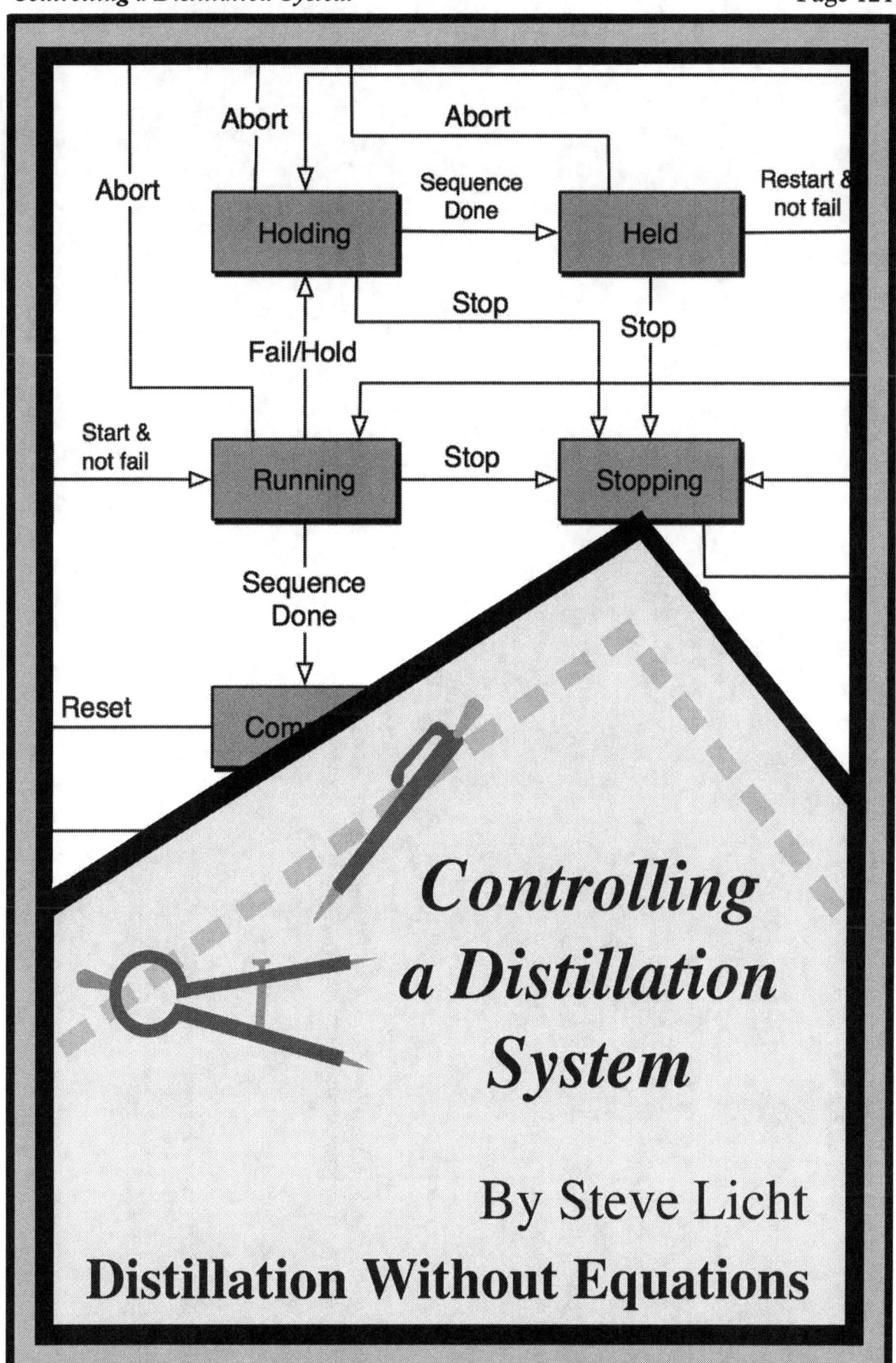

Controlling a Distillation System

By Steve Licht

Distillation Without Equations

Controlling a Distillation System

Part of the

Distillation Without Equations
Technical Seminar

By Steven Licht

Distillation process control: Controlling what you are distilling

- Many levels of control simplicity or complexity are possible
- Each has its merits & is appropriate someplace
- First understand what goes on inside a distillation system that requires control, before determining how simple or complex your system control scheme & control hardware should be
- Knowing how to start up, line out, shutdown, & handle emergencies helps to provide a robust control scheme
- Consider the diverse types of field instrumentation available, the types of process controllers available, & their merits & limitations, before selecting control strategies & specifying them on P&ID's & in instrument lists
- Computerized PLC & DCS system hardware & software design presents additional challenges
- Full automation is possible, following ISA S88 model for batch plant control, or simpler sequential function chart logic

What requires control?

A control strategy (manual or automatic) is required for the following:

- Feed input
- Heat input
- Column product purity control
- Column bottom effluent
- Reflux
- Column top distillate
- Other feed, product, or internal streams
- System pressure

Manually or automatically set control valves carry out the selected control strategy

What needs monitoring & recording?

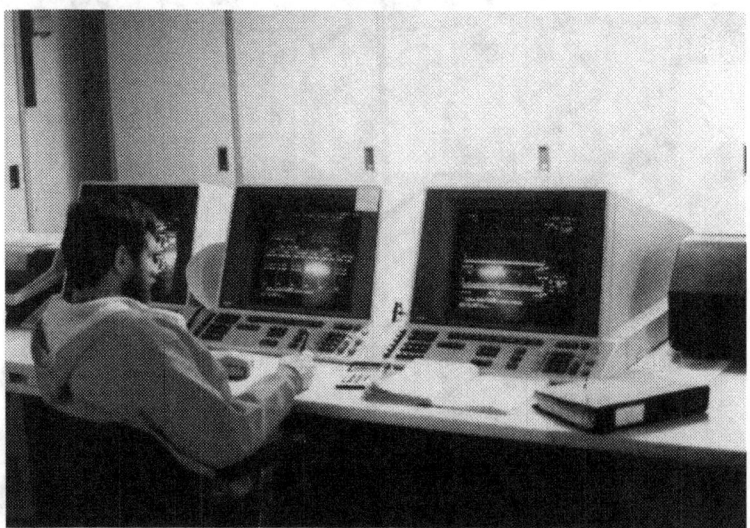

In addition to control loops, process indicators & safety interlocks require monitoring & recording

- Indicators are for operator information or for data recording purposes
- Interlock devices monitor their inputs & respond if required for protection of pumps & overall system safety

What other things are standing by, just in case they are needed?

Safety equipment is standing by:

- Pressure safety valve (PSV) or burst disk pressure safety element (PSE) relieves pressure in each section of the system containing a vessel that can be blocked in
- High level switches & low level switches
- High pressure switches
- Interlocks triggered by switches or alarms

Let's discuss what is going on inside this distillation system, before we learn to stop or start it, or how to handle problems

- What is turned on?
- What is flowing?
- What is heating?
- What is boiling?
- What is condensing?
- What is cooling?
- What is contacting?
- What is pumping?
- What is coming in & going out?
- What requires control?
- What needs monitoring & recording?
- What needs sampling?
- What other things are standing by, just in case?

What is flowing?

- A hot fluid is providing heat
- A cold fluid is providing cooling
- Feed material is flowing in
- Purified distilled product is flowing out
- Waste streams are flowing out
- Column bottoms are being boiled
- Column overheads are being condensed
- Part of the condensed overhead flows back to the top of the column as liquid reflux
- Gases that cannot be condensed are flowing out

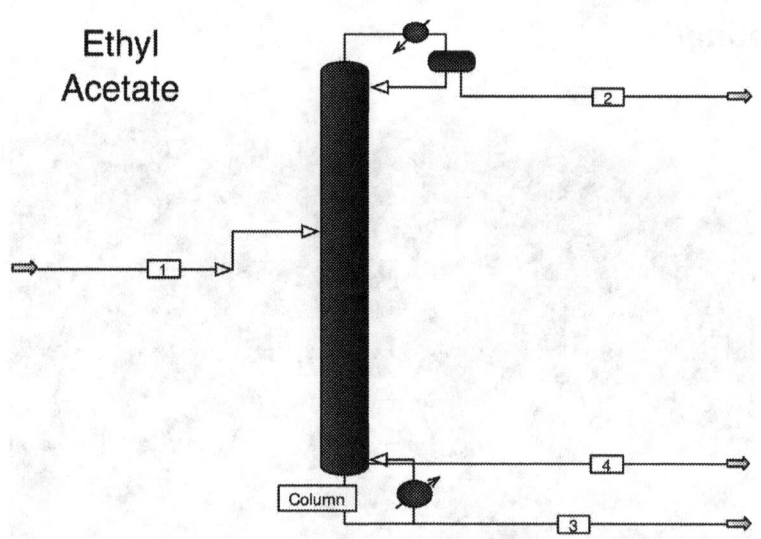

Ethyl
Acetate

Ethyl Acetate system flow: 1: Feed 4: Product 2,3: Wastes

Stream Name	1	2	4	3
Stream Description	Feed	OVHDS Waste	Main Product	CRUD
Phase	Liquid	Liquid	Vapor	Liquid
KG/HR	2310.000	1751.617	548.383	10.000
Temperature C	10.000	64.721	82.587	82.910
Pressure BAR	4.000	1.000	1.209	1.220
Molecular Wight	69.989	65.710	87.950	88.086
Weight Comp. Percents				
H2O	4.3290	5.7033	0.0182	0.0007
Ethanol	6.2771	8.2564	0.0689	0.0093
EOAC	72.6407	63.9912	99.7706	99.9513
IPA	0.5195	0.6659	0.0611	0.0146
MEAC	16.2338	21.3832	0.0811	0.0241
Weight Comp. Rates KG/HR				
H2O	100.0000	99.9000	0.1000	0.0001
Ethanol	145.0000	144.6210	0.3780	0.0009
EOAC	1677.9999	1120.8801	547.1246	9.9951
IPA	12.0000	11.6632	0.3353	0.0015
MEAC	374.9999	374.5526	0.4450	0.0024
Enthalpy M*KCAL/HR	0.031	0.077	0.069	0.000

What is boiling?

Ethyl Acetate system flows: Mass & heat, in & out, at T & P

Column Summery						
Tray	Temperature C	Pressure BAR	Net Flow Rates			Duties
			Liquid	Vapor	Feed	Product	
			KG-MOL/HR				K*KCAL/HR
1	64.7	1.00	179.2			26.7	-1.5920
2	68.2	1.02	181.0	205.8			
3	69.0	1.03	181.1	207.6			
4	69.6	1.04	181.3	207.7			
5	70.1	1.05	181.5	207.9			
6	70.5	1.06	181.8	208.2			
7	70.9	1.07	182.1	208.5			
8	71.2	1.08	182.4	208.8			
9	71.5	1.09	225.2	209.1	33.0		
10	72.0	1.10	225.8	218.9			
11	72.4	1.11	226.4	219.4			
12	72.8	1.13	227.0	220.0			
13	73.1	1.14	227.7	220.7			
14	73.5	1.15	227.8	221.3			
15	74.1	1.16	225.9	221.5			
16	75.9	1.17	222.3	219.5			
17	79.2	1.18	222.8	216.0			
18	81.3	1.19	224.2	216.4			
19	83.1	1.20	224.9	217.9			
20	82.6	1.21	225.3	218.6		6.2	
21	82.9	1.22		225.1		0.1	1.7076

What is contacting?

Ethyl Acetate column internal flows: Liquid & vapor by stage

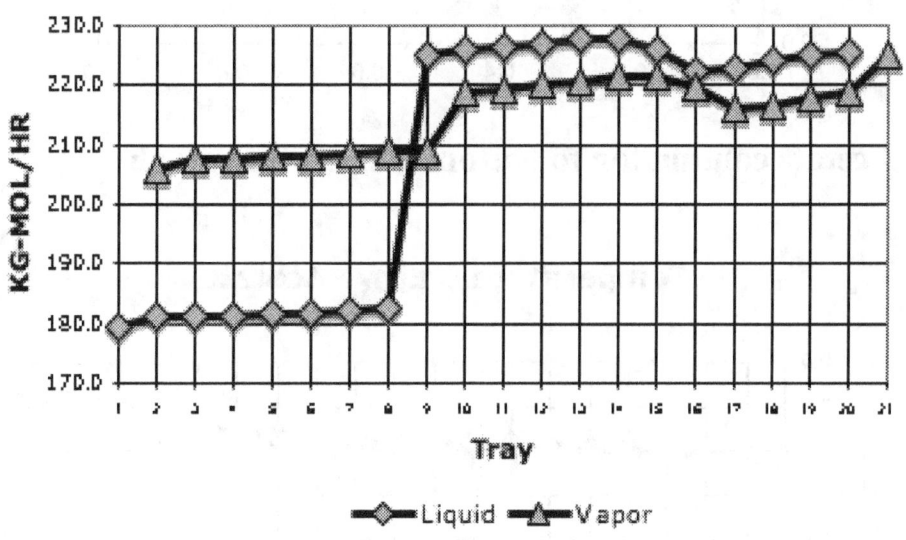

Rates Ethyl-Acetate

Ethyl Acetate distillation column liquid & vapor flows, all trays

The purity changes stage-by-stage as vapor contacts liquid

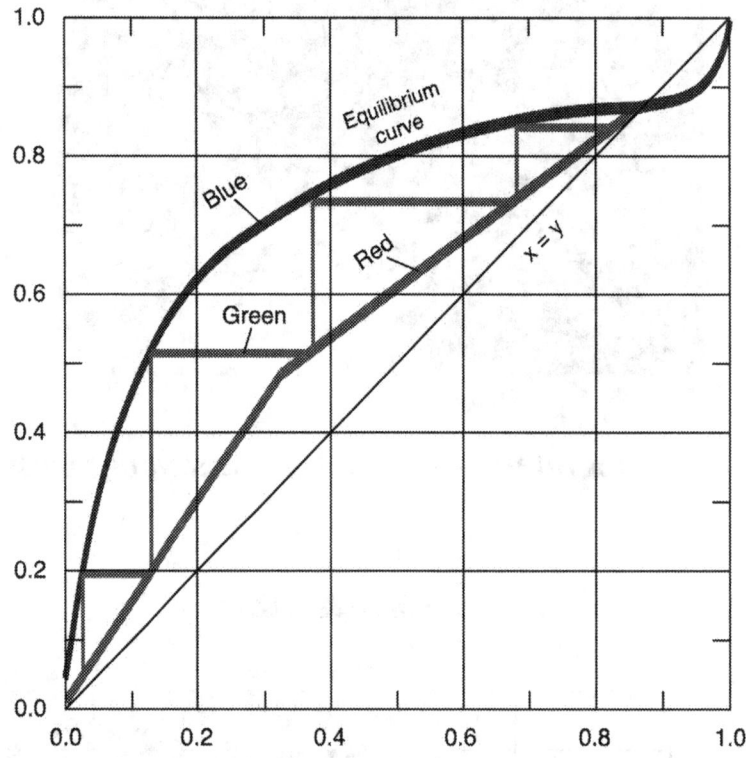

Ethyl Acetate column top to bottom temperature graph

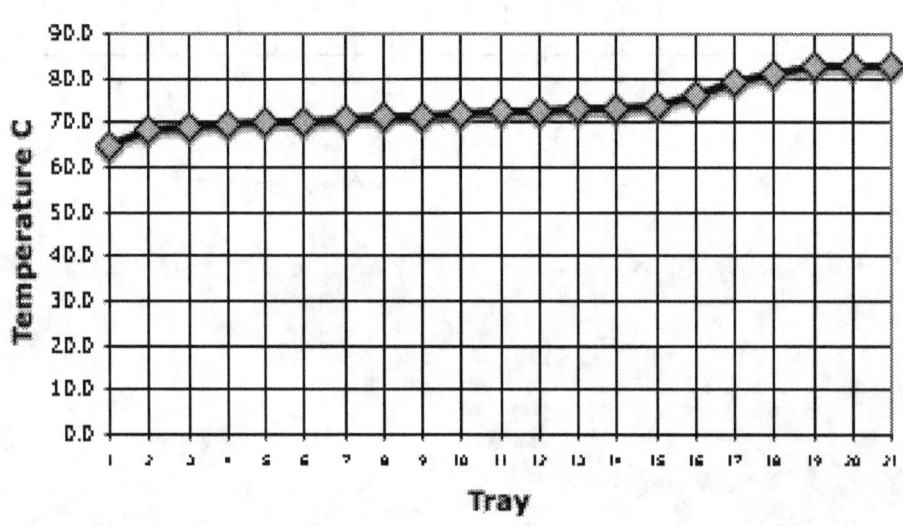

Ethyl Acetate column temperature profile

Deck of a valve tray with round valve caps, two pass liquid path

Underside of a valve tray with round valves & downcomers

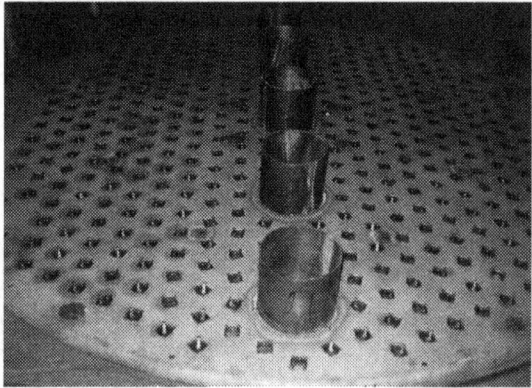

Trays are mounted horizontally inside distillation columns

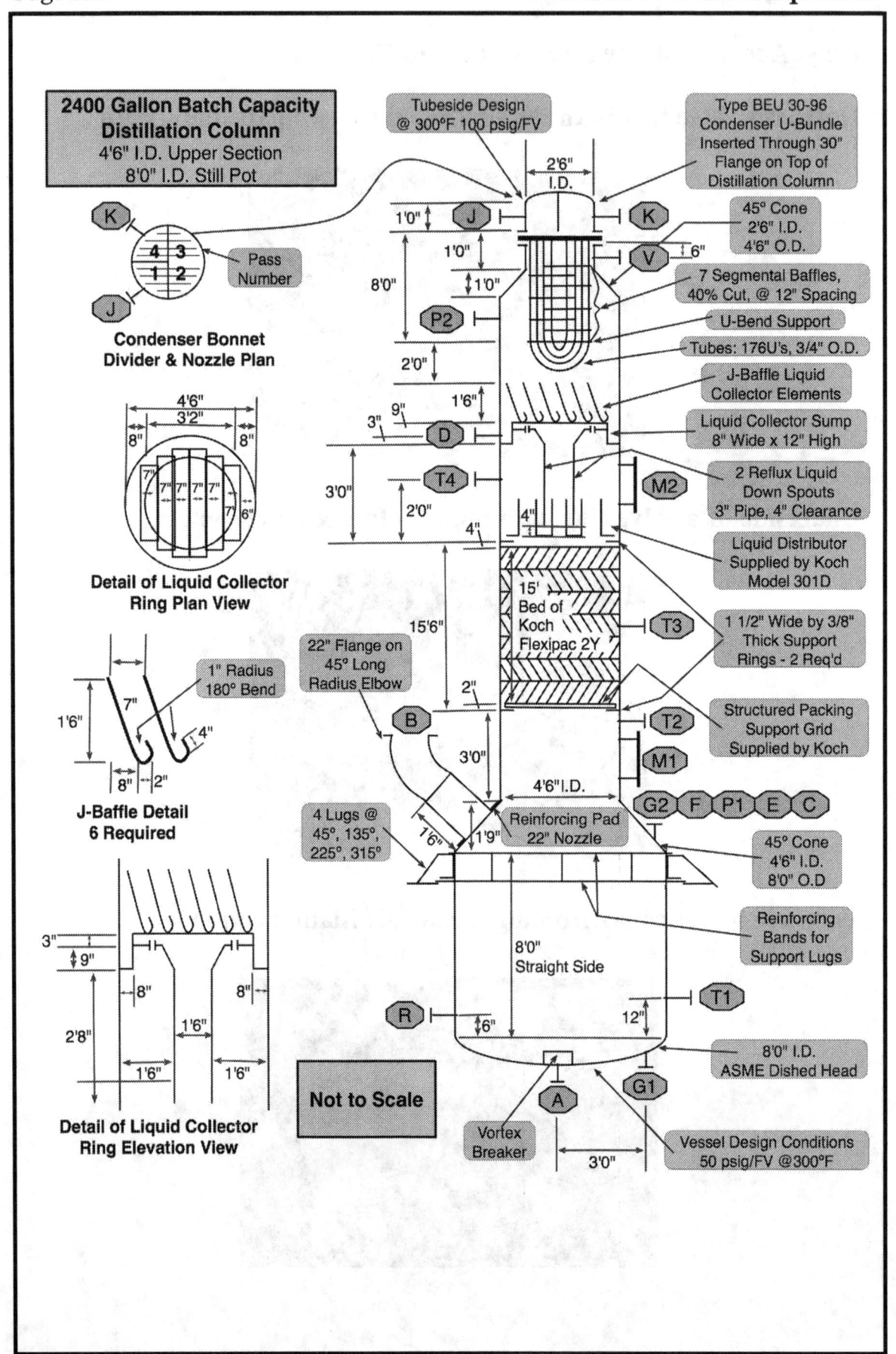

**2400 Gallon Batch Capacity
Distillation Column**
4'6" I.D. Upper Section
8'0" I.D. Still Pot

K

4 3
1 2

Pass
Number

J

**Condenser Bonnet
Divider & Nozzle Plan**

4'6"
3'2"
8"   8"
7" 7" 7" 7"
7"   6"

**Detail of Liquid Collector
Ring Plan View**

1" Radius
180° Bend
7"
1'6"
4"
8"   2"

**J-Baffle Detail
6 Required**

3"
9"
8"   8"
2'8"
1'6"
1'6"   1'6"

**Detail of Liquid Collector
Ring Elevation View**

Tubeside Design
@ 300°F 100 psig/FV

2'6"
I.D.

1'0"   J

1'0"

1'0"

8'0"

2'0"

P2

3" 9"   1'6"

D

T4   3'0"

2'0"

4"   4"

15'
Bed of
Koch
Flexipac 2Y

15'6"

22" Flange on
45° Long
Radius Elbow

2"

B

3'0"

4 Lugs @
45°, 135°,
225°, 315°

1'6"   1'9"

Reinforcing Pad
22" Nozzle

4'6"I.D.

8'0"
Straight Side

R

6"

12"   T1

Not to Scale

Vortex
Breaker

A   3'0"   G1

K

Type BEU 30-96
Condenser U-Bundle
Inserted Through 30"
Flange on Top of
Distillation Column

45° Cone
2'6" I.D.
4'6" O.D.
V   6"

7 Segmental Baffles,
40% Cut, @ 12" Spacing

U-Bend Support

Tubes: 176U's, 3/4" O.D.

J-Baffle Liquid
Collector Elements

Liquid Collector Sump
8" Wide x 12" High

M2   2 Reflux Liquid
Down Spouts
3" Pipe, 4" Clearance

Liquid Distributor
Supplied by Koch
Model 301D

T3   1 1/2" Wide by 3/8"
Thick Support
Rings - 2 Req'd

Structured Packing
Support Grid
Supplied by Koch

T2

M1

G2 F P1 E C

45° Cone
4'6" I.D.
8'0" O.D

Reinforcing
Bands for
Support Lugs

8'0" I.D.
ASME Dished Head

Vessel Design Conditions
50 psig/FV @300°F

What is pumping?

There are different types of pumps for different duties

Positive displacement pumps:

- Cannot have flow restricted, while they are running - Never
- Can maintain a fixed flow rate, despite variations in back pressure
- Can be much easier to prime than centrifugal pumps
- Useful for transfer pumps that must be frequently drained

Centrifugal pumps:

- Can vary in flow rate across a wide range
- Any change in back pressure changes the flow rate
- Can be cut back to zero flow for approximately one minute without damage
- One hour at zero flow will overheat & damage a centrifugal pump

What is coming in?

IPA distillation 1: Feed 2: Waste 3: Product

IPA/Water			
Stream ID	1	2	3
Temperature (C)	15.0	81.9	82.8
Mass Flow (KG/HR)	800.000	150.924	649.079
Mass Fraction			
Water	0.002	0.010	100 PPM
ISOPR-01	0.998	0.990	1.000
METHA-01			
DIBUT-01			

What is going out?

IPA / Methanol distillation system 1: Feed 2,4: Waste 3: Product

IPA/Methanol/Water				
Stream ID	1	2	3	4
Temperature (C)	15.0	66.3	82.8	74.6
Mass Flow (KG/HR)	500.000	27.000	446.000	27.000
Mass Fraction				
Water	0.002	0.008	102 PPM	0.028
ISOPR-01	0.948	0.249	1.000	0.790
METHA-01	0.050	0.744	918 PPB	0.182
DIBUT-01				

What electrical devices, utilities, & equipment are turned on?

- Control panel power
- Pump motor power
- Agitator motor power
- Instrument air
- Cooling source
- Heating source
- Fire protection
- Phone, radio, security

Control panel devices requiring power supply turned on:

- Panel AC & DC
- UPS input & output
- PLC or DCS
- HMI or SCADA
- PC operator stations
- Alarm panels
- Analyzers
- Printers

Motor Control Center (MCC) devices turned on:

- MCC 3-phase supply
- Motor starters for all motors in use
- Breakers to instrument power transformers
- Breakers to lighting panels
- Breakers to UPS feeds

Pumps & agitator motors switched on - possible ways:

- On/off switch at MCC
- Local on/off switch by motor or on panel
- Local hand/off/auto switch by motor or in panel
- Automatic run from control panel relay logic
- Manual run from HMI
- Automatic run from PLC or DCS system

Compressed air supply for instruments turned on:

- Instrument air compressors running
- Instrument air dryer working
- Air supply pressure OK
- Instrument air filters OK
- Air valves on headers & branch lines to instruments are open
- Air bleed & drain valves are closed

How to fill an empty distillation system to prepare for start up

- First make certain that no one is still working on or inside of equipment
- Make certain that all equipment & piping is in place, blinds are removed, drains are closed, & normally open manual valves are open
- Only the distillation column base & reboiler circulation loop need to be filled to normal liquid levels before a start up
- Other equipment such as the decanter & reflux drum can remain empty at this time

How to start up a cold (but not empty) distillation system

- Make certain that all cooling systems & vent collection systems are working
- With a forced circulation reboiler, start the reboiler circulation pump
- Slowly ramp open the heat source to the reboiler (steam or hot oil) to near normal heat input
- When material begins boiling & the column level decreases, add more feed to top up the column bottom as required, whenever required
- When the column is hot all the way to the top, condensate will appear in the reflux equipment

How to line out a newly started or restarted distillation system to achieve a total reflux condition

- When condensate appears in the reflux equipment, start the reflux pump, allowing all of the reflux to return to the column
- Keep topping up with feed material as needed
- Ramp the heat input rate up to normal
- When column temperatures line out & no more material top ups are required, the system is ready to begin accepting continuous feed

How to start feed, recycle, & line out a distillation system to normal compositions & flows

- Ramp up feed rate quickly to 75% of normal
- Make certain that control loops for bottoms flow out & distillate out are functioning
- Frequently bottoms & distillate are recycled back to the feed tank until on-spec
- When column temperatures stabilize, increase feed to 100% of normal, & allow to re-stabilize

When & how to recognize that a distillation system has begun to make good product

- All the flow rates in the system will be near their usual values
- All the column temperatures will be near their usual values
- A trend recording of the temperatures & flows will show straight lines
- On-line analyzers & density meters will show values within good quality specifications

How to continue long-term operation with on-spec quality

- Primarily, it was the designers' duty to make certain that the distillation system includes robust control configurations which ensure quality
- If you must control something manually to maintain on-spec quality, increasing the amount of reflux usually improves the distillate quality & increasing the heat input usually improves the bottoms stream quality
- Decreasing the feed rate improves quality at the top & bottom of the column, in most systems
- Flow blockages often show up as quality crashes

What to do when quality measurements fail

- First, resample & reanalyze, to be sure that the quality failed, not the measurement
- If uncertain about what to do to improve quality, decrease the feed rate by 25% as a first step
- If quality improves, gradually bring the feed rate back up towards the normal value
- If quality did not improve, increase the reflux next
- If quality still did not improve, increase the heat input, watching out for the signs of column flooding (high pressure drop across column)
- If nothing you do works, call someone to help out

What to do about utility variations:

- Quality
- Quantity
- Supply Pressure
- Temperature
- Unsteadiness
- Cycling

Problems that might come along - what to watch for & what to do

Figure 3

Pump stuffing box leakage, probable electrical insulation failure and air in this pump room completed the fire triangle with explosive force. Adequate ventilation and well-maintained equipment will eliminate such accidents.

Figure 21

Breakage of a glass enclosed "look box" released light naphtha. Vapors spread to a hot tar line where they ignited and flashed back, exploding other combustible mixtures in the area.

Emergency handling: Intervention & shutdown

- If safety interlocks or emergency systems are triggered, do not prevent them from acting
- When there is an emergency in a system equipped with emergency stop (E-Stop) buttons, use them
- If required to intervene manually, first block any energy inputs to the system (steam, hot oil), then shut off all electric motors, then block all material inputs to the system
- It is usually better to let pressure safety valves & burst disks relieve system pressure, than to open vent valves to atmosphere during an emergency

Figure 4
Fire at refinery storage area

How to make a smooth shutdown

- Ramp the feed flow rate down to zero
- Ramp the heat input flow rate down to zero
- Keep all other instruments functioning, control loops in automatic, pumps running, & utilities flowing
- As material output flows decrease naturally, close their flow paths & stop pumps
- When all pumps have stopped, & vessels contain only safe residual liquid heels, vent or inert the system as prescribed for shutdown periods

When & why to empty a distillation system

- A distillation system should be emptied when the next production material is not compatible with the liquid heel left inside after a shutdown
- Some or all of the equipment or piping may need to be emptied & blinded off to make it safe for work to be done during the shutdown period
- Drain all pipelines to vessels or to drain connections, pump all vessels empty, blow out lines with nitrogen or steam, water rinse, air dry
- Do not work inside nitrogen blanketed equipment

Prerequisites to preparing a new distillation system control strategy

- Steady state material balance must exist
- Equipment process duties & connection sizes must be known
- Consider how to fill the system, start it, shut it down, & empty it
- What valves, instrumentation are required to operate this plant safely?
- What control scheme will be keep this system going, safely making good product day after day?

Instruments in a distillation system may be categorized as follows:

- Sensors touch the process & "feel" or sense conditions
- Transducers change this "sensation" into a standard signal that can be transmitted elsewhere, either by pneumatic air pressure, milliamp current, voltage, HART digital protocol added on top of milliamp signals, or directly digital field buses
- Indicators give a readout for you to see & understand
- Controllers actively respond to process & setpoint changes, trying to maintain the process measured variable equal to its setpoint value
- Control Valves move as directed by the control system to change flow characteristics as required to modify the process measured variable

Field Instrumentation

- How field instrumentation works can be difficult to understand, because there are so many different types & designs
- However, most of the action & most of the problems occur in the field instruments, not at the control panel, so putting some effort into understanding how the field instruments really work & what they are really sensing is worthwhile

Field Instrumentation Categorization

Pressure	Direct Readout	Bourdon Tube Gauge Manometer
Pressure	Transmitted Signal	Diaphragm Bellows Solid State Cell
Temperature	Direct Mechanical	Liquid-in-Glass Bimetal (Dial) Capillary Tube
Temperature	Electrical Signal	Thermocouple Resistance (RTD) Thermister
Liquid Level	Direct Readout	Sight/Gauge Glass Float & Rod Float & Cable
Liquid Level	Head Pressure	Diaphragm Bubbler
Liquid Level	Electrical	Capacitance (RF) Vibration Ultrasound, Radar
Other	Misc. Fluid Properties	pH, Conductivity Specific Gravity Viscosity
Flow Rate	Differential Pressure	Orifice Plate, Wedge, Venturi, Flow Nozzle, Pitot Tube, Annubar, Elbow Meter, Target Meter
Flow Rate	Variable Area	Rotameters, Variable Area Flow Tubes
Flow Rate	Positive Displacement	Nutating Disk, Vane, Gear or Lobe, Turbine
Flow Rate	Principles of Advanced Physics	Vortex Shedder, Electro-magnetic, Ultrasound (time-of-travel), Ultrasound (Doppler), Mass (Coriolis), Mass (thermal)
Flow Rate	Open Channel Flow	Weir (V-Notch), Flume

Field Control Instruments

Control Inst.	Local Mechanical	Pressure Regulator, Temperature Control, Level Control
Control Inst.	Remote Pneumatic	Transmitters, Controllers
Control Inst.	Remote Electrical	Transmitters, P/I Transducers, E/I Transducers, I/P Converters, Field Bus Interfaces, Indicators, Controllers
Discrete Inst.	Process Switches	Pressure, Temperature, Position, Level, Low Flow

Control Panel Instruments

Indicators & Recorders	Pneumatic Electronic Digital computer based
Controllers & Recorders	Pneumatic Electronic Digital computer based
Alarm Systems	Electrical (Relays) Electronic Digital computer based

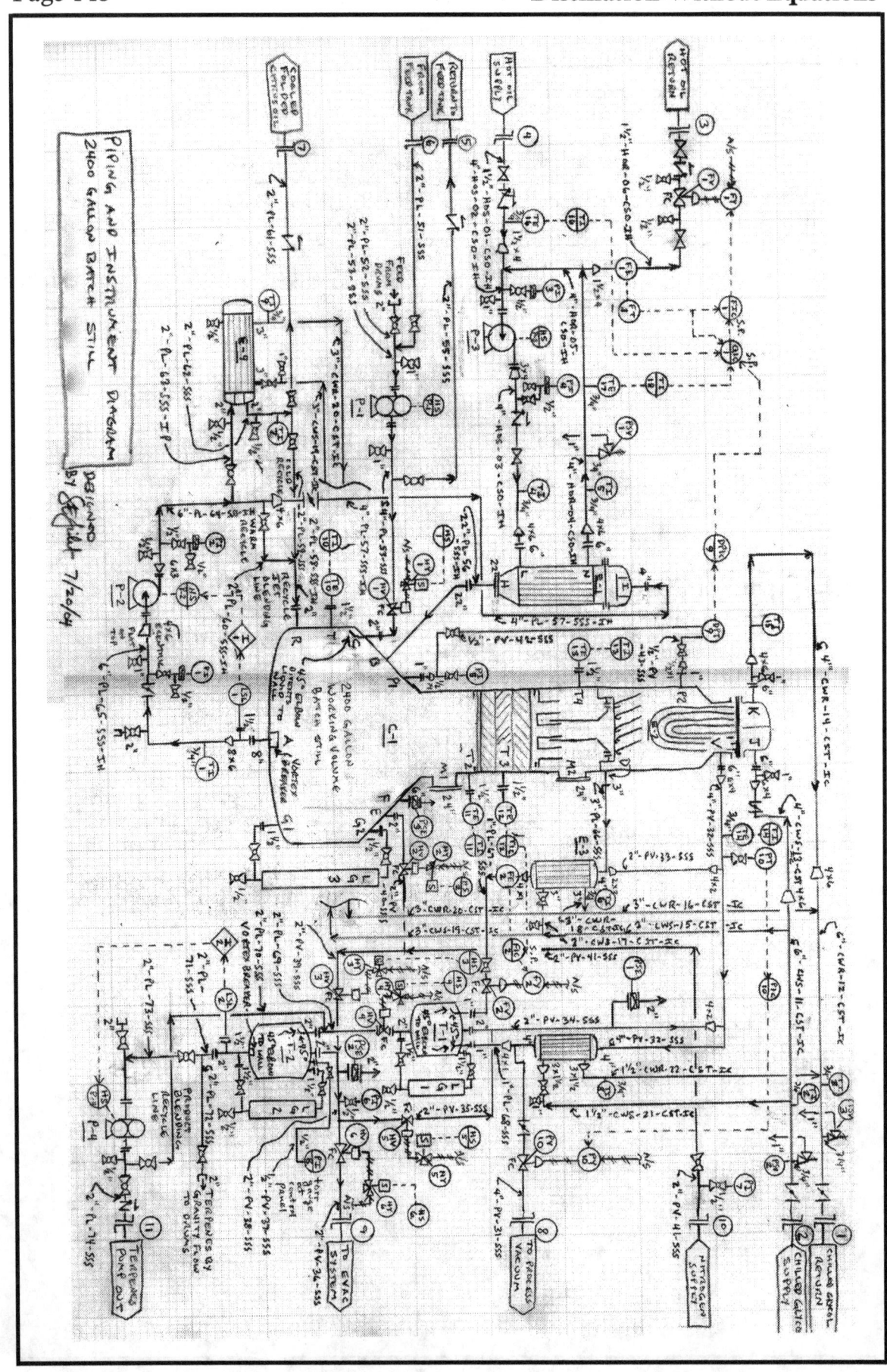

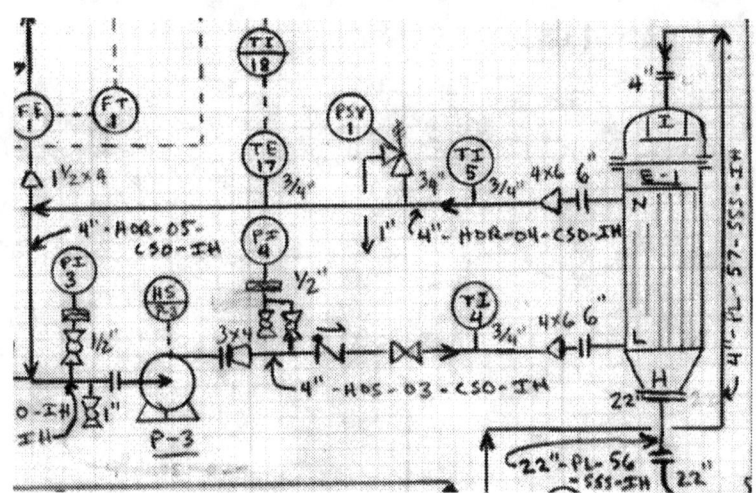

Process control

- Control of processes can be accomplished many different ways
- Some of these ways are safer, more accurate, or more automatic than others
- Manually controlled distillation systems are still best for simplicity & long term reliability, especially with unsophisticated operations & maintenance staff
- The more advanced automation that can be implemented on PLC & DCS platforms is only justified in very high production environments

Types of process control

- Uncontrolled or "wild", like city water temperature
- Manual control, like a room heated by wood stove
- On-off control, like a room heated by furnace with thermostat
- Feedback control, like a car cruise control
- Feed forward control, like a multi-stream product blending system
- Advanced control modes, like multi-fuel boiler combustion controls
- Most common are manual, on-off, & feedback control

Typical feedback control loop

- Applies to control of temperature, pressure, liquid level, or flow
- Can be Proportional (P), Proportional Integral (PI), or Proportional Integral Derivative (PID)
- Loop dynamics will depend on the settings employed for gain, reset time, & derivative time
- Loop stability will depend on how well these settings suit the natural process value ranges (as compared to calibrated instrument ranges), the speed of changes, the length of time lags, & amplitude of noise interfering with measurement

Feedback flow control loop

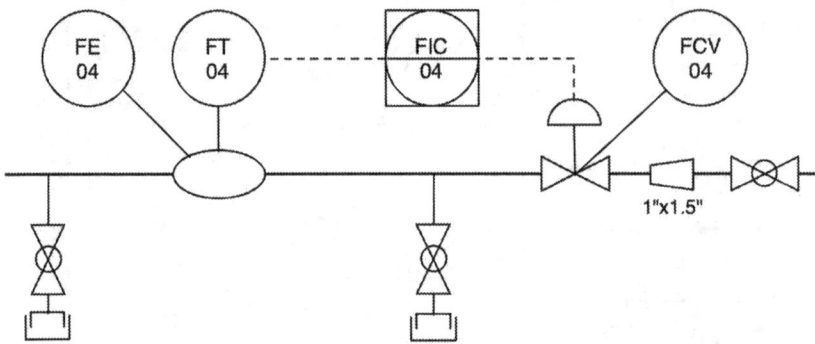

Switching between manual & automatic process control

- In manual mode, you control the output (that is, open or close control valve) directly from the control panel or computer interface screen
- In automatic mode, you adjust the setpoint only, & the controller controls the output (that is, opens or closes the control valve) to keep the process steady
- Some DCS systems use different terms:
  - What is called "automatic" here is called "cascade" there, & what is "manual" here is "automatic" there!

Stable processes

- Stability of a process under control is essential to safe operation of the equipment being controlled & the whole plant facility
- Stable Processes stay within safe bounds of temperature, pressure, flow, & liquid level
- Unstable processes can get too hot or too cold, be under or over pressurized, flow too fast or too little, run tanks empty or overflow them
- All processes are open loop stable without controllers - they will steady out by themselves
- All processes can be made closed loop unstable, if poorly designed or poorly tuned controllers are added!

Cascaded feedback control loops add stability to second order or long time lag processes

- Cascaded feedback loops use the output of one loop (usually level or temperature control) as the setpoint for another loop (usually flow control) thereby isolating the primary control loop from disturbances occurring in the second control loop

The same control loops can be implemented on many different hardware platforms:

- Pneumatic controllers with 3-15 psig I/O air signals to & from the field
- Electronic controllers, single loop or multi-loop, 4-20 ma or other AC or DC I/O signals to & from the field
- Microprocessor computer based, PLC or DCS, electronic signals to & from the field
- PLC or DCS with digital signals added to analog signals to & from the field (HART), or digital bus communication (Field Bus, Asi Bus, etc.)

More complicated control concepts also fit onto pneumatic & electronic platforms:

- Ratio control
- Mathematical operation blocks such as square root extraction, addition, multiplication, Boolean logic
- Cascade, output of one controller used as the setpoint for another controller
- Fuzzy logic:
    - Multiple inputs or outputs used for one process control application

Sequencing of actions is possible when microprocessor based PLC or DCS systems are used

- The same control loops that can be implemented on pneumatic or electronic stand alone controllers can be implemented on PLC or DCS platforms
- Sequential function charts can be provided to cause actions such as start up, shut down, emergency actions, & operating sequences to occur automatically
- Batch plant processing concepts can be implemented in PLC systems & DCS systems, usually following the ISA S88 standard

Data collection & storage is available with PLC & DCS systems, or separate data historians

- DCS systems have built in data historians
- PLC systems usually provide data historian capabilities within the SCADA system or via other linked software
- Data collection & storage from electronic or pneumatic controllers require transducers to digital protocols, similar to PLC & DCS I/O cards
- Stand-alone pneumatic & electronic controllers probably should be replaced with PLC or DCS systems when large electronic data collection & storage requirements are desired

Relative merits & drawbacks of alternative distillation process control schemes

A control strategy (manual or automatic) is required for the following:

- Feed input
- Heat input
- Column product purity control
- Column bottom effluent
- Reflux
- Column top distillate
- Other feed, product, or internal streams
- System pressure

Some general truths about what you can & cannot control in a distillation process

- There is one parameter that always must be controlled:
  - System pressure as measured at one point, usually in the column head space or condenser vent line
- The number of other different things that you can independently control in a distillation process usually is equal to the number of streams entering the process, plus the number of streams leaving; reflux does not count
- Typically, you try to control one thing about each stream:
  - Either control its purity, control its flow rate, or alter the liquid level in a vessel where this material accumulates thereby controlling inventory in the system
- If you try to control two things about any one stream, then you must accept that you will control nothing about one other stream in exchange for this privilege; if you control reflux independently, you surrender something else
- It is most common to control the feed flow rate, the heat input rate, & the purity or flow rate of all except one of the column effluent streams - top, bottom, or side draw

History suggests that simplicity is better than complexity, that successes be repeated, & that failures not be repeated - hence, these guidelines

- No control scheme works all of the time, but these guidelines will try to make suggestions that will work most of the time
- First think about how you would manually control any distillation system you design, if the only instruments available were dial thermometers, visual sight level indicators, rotameters or visual "look box" flow indicators, specific gravity measurement floats, & manual globe valves
- In the undeveloped regions populated by more than one half of the world's people today, that is probably how the system should be controlled
- The developed world's more sophisticated control systems only function by doing things that people could do manually if they were constantly vigilant, well trained, & did not mind cold or heat or high places
- Remember that the 18th century European armies & navies received rum, brandy, or vodka in their rations, so someone must have been able to distill the rum, brandy, or vodka using only 18th century control technology!
- To quote Albert Einstein:
    - "Everything should be made as simple as possible, but not simpler"

Control of reflux flow to the distillation column

- During startup, total reflux is usually required, so the controls must allow for return of 100% of the condensed overhead vapor as reflux
- Either the reflux must return to the column naturally by gravity, or reflux accumulator level control is required to pump all the liquid back as reflux
- Usually it is best from the control point of view to keep reflux under some sort of natural level control, manipulating the distillate product stream flow based on temperature control for purity, with the balance of the condensate returning as reflux
- Remember that condensed vapor rate minus reflux equals distillate product rate, so if you choose to control the reflux rate independently, you lose the opportunity to control the distillate flow
- In some systems where distillate purity is sensitive to top temperature, controlling the top purity by temperature control cascaded to reflux flow control is a good scheme - but this might not get you through startup
- Trying to control reflux by its ratio to the distillate flow rarely yields stable control, except during very smooth & steady operation periods
- A system with zero reflux loses one of its other independent control opportunities

Control of partial condenser vapor product rate or composition

- You may not think that your system condenser is running as a partial condenser, but it nearly always is
- Fixed gases dissolved in the feed materials, & low boilers under the column top conditions, largely pass through any condenser provided
- Some distillation systems vent a significant amount of gas through their partial condensers, some less, but rarely is the vent flow equal to zero
- It is difficult or impossible to control the vent vapor stream from a partial condenser by measurement either of its temperature or its flow rate
- What is most practical to control, is the temperature of the coolant inside the condenser
- The vent vapor temperature nearly always is very close to the coolant temperature
- If utilities are not available at the desired temperature, or if the coolant temperature must be adjustable by the process, use a tempered coolant recirculation loop

Control of an overhead waste stream as a decanted second phase

- When the overhead condensate stream forms two or three liquid phases, one being aqueous & the others organic, it is almost always desirable to let the phases settle by gravity, & let the aqueous phase go a different way than the organic
- Letting the phases enter a decanter vessel under very calm internal conditions usually is sufficient for practical success in settling
- The lower the difference in specific gravity between the phases, the lower the settling velocity of the mixed phase materials, so low, wide decanters work more quickly than tall thin ones
- External or internal manometer principle or DP control loops will succeed at keeping the interface within a controlled range, if both phases are of known unchanging specific gravity
- If different specific gravities must be accommodated, a device which detects the physical location of the interface, such as a capacitance probe or weighted magnetic float, will work better than DP or manometer loops
- Part of the reason why organic phases do not separate from one another so easily as an aqueous phase separates from an organic phase, is that their surface tensions are nearly equal, therefore the organic phases may form continuous intermingled sheets (like wood grain) rather than discrete drops which readily rise or fall & merge when they touch

Distillate product purity & flow control

- Whether the distillate product is a desirable product, or instead is a waste stream, it is still most common to control its composition by control of temperature somewhere in the upper part of the column, rather than rely solely on a column bottoms product composition control scheme

- The reason it is more common to control the top product purity than the bottom, is that the top of the column is usually much quicker to respond to a flow change with a measurable temperature change, because of fewer time lag opportunities
- It is usually best to control distillate purity by temperature control from a point inside the distillation column, near the top, but in a sufficiently impure zone that the normal tray compositions at this point display a pleasantly large & predictable change in temperature between the stages
- The column temperature controller output can be used to control the distillate valve directly, but it is far better to cascade the temperature controller output to the setpoint of a distillate flow control loop, providing a degree of damping that greatly helps stabilize the reflux flow & thus the composition & temperature profile in the column
- When the reflux flow is fixed by flow control without reset from decanter or reflux drum level, or is cascaded from column temperature control, then the distillate flow cannot be independently controlled, & must be driven by reflux drum level

Temperature control setpoint pressure correction method

- Usually when distillation column temperature is being controlled, it is for the purpose of controlling composition of an effluent stream
- The temperature setpoint value at the control point usually can be expressed as a difference between the temperature measured at this point, & the boiling point temperature of the pure component which dominates the column inventory at this point
- The actual temperature experienced inside the column shifts with column pressure & composition
- The desired temperature difference between the control point temperature & the pure component boiling temperature is a function of composition only, & does not shift with pressure
- The effects of varying pressure on measured temperature can be filtered out of the purity control scheme by taking the system pressure at the temperature measurement point, possibly calculated by lever arm rule from other pressure measurement points, & then using Antione's equation for vapor pressure to calculate the pure component boiling temperature at system pressure
- Control is then based on temperature difference from the calculated pure boiling point, rather than on absolute pressure measurement
- This scheme has worked in practice, in diverse applications, time & again

Control of feed flow into a distillation system

- A fixed feed flow rate is usually best for continuous distillation systems, as it is for most production line operations
- The feed flow rate may be changed either continuously or in step changes at discrete points in time, to match the distillation system processing rate with the production rate of upstream processes
- A distillation system will usually keep working well through step changes in feed rate or feed composition, so long as the mass flow of the most difficult impurity to remove is kept below a fixed maximum rate, & the mass flow of the primary product to be purified is also kept below a fixed maximum rate
- Temperature control at some point in the distillation column may be cascaded to control of the feed flow rate, but this loop would have a large inherent time lag, & thus would be apt to cause cycling
- If the bottoms level is used to control feed flow, a large process time lag is likely to combine with a very insensitive process variable (bottoms level) to spawn a bounded cycle in the value of most distillation system flows & compositions
- Zero feed flow is normal for a batch distillation system, but be aware that with zero feed, all of the parameters of the distillation system operation (inventory, level, temperature, purity) will vary over time

Control of a second feed stream going into a distillation system

- A second feed stream may mean that there are multiple feeds into the column containing similar components, or the second feed may be very different in composition, as in extractive distillation
- If the second feed stream is just an alternate feed containing the same primary product as the first stream, then either or both feed streams may be varied independently, so long as the rules about maximum impurity mass flow rate & maximum primary product mass flow rate are respected
- If the second feed stream is an extractive distillation agent, it should be flow controlled, with its setpoint value determined by a ratio flow controller which watches the first feed stream flow rate

Control of distillation process heat input

- Distillation systems operate with the best overall stability when their heat input rate is steady
- When the heat input comes from steam heating, then it is best to control the flow rate of steam going to the reboiler
- When hot oil or other sensible hot fluid is used for heating, then the flow of this liquid must be controlled, the temperature of the heating media in & out must be measured, the temperature difference must be multiplied times

the flow rate to determine the heat input, & the flow rate setpoint must be adjusted to maintain constant heat input

- Heat input control in Mechanical Vapor Recompression (MVR) systems usually is achieved by controlling the motor speed or inlet vane position of the MVR compressor
- Heat input control for multiple effect column trains, where one column heats another column, can only be done at the first effect column, where the fresh heat source is - & it is very crucial that the heat input be very steady & stable, because there are high order time lags inherent to such a design
- Occasionally it is appropriate to use column lower zone temperature controller output or bottoms level controller output, cascaded to the heat input flow controller to trim the heat input rate, but such a loop must be low gain, primarily integral, with a experience-tuned derivative feature useful to ensure the stability of this loop when the feed flow or feed composition changes

Control of column differential pressure

- Column differential pressure is often used as a control parameter by design engineers
- I have yet to see any column differential pressure control working in automatic mode without it causing continual cycling of the column flows, temperatures, & purity
- The reason for this cycling, is that the column pressure drop is a function of multiple independent variables
- Column differential pressure naturally varies slowly over time in a well run distillation system, in response to changes in liquid traffic & vapor traffic up & down the column, which themselves may be responses to changes in feed composition or other uncontrolled parameters
- Control engineers like to set heat input rate by control of column differential pressure, but an empty distillation column never can be started up using such a control feature, because the relationship between heat input rate & the column pressure drop before it is fully loaded with liquid & vapor traffic is constantly changing
- Load changes made while lining the system out tend to start cycles which never seem to go away, if the column differential pressure is used to control heat input or feed rate
- After a large column is well lined out, differential pressure control may be used in a slow insensitive PI loop, to trim either the heat input or in some cases the liquid feed rate
- In very low pressure operations with packing, where the measured differential pressure is indeed well related to vapor flow rate, differential pressure control in a low gain PI loop may be used to adjust the heat input in order to keep the packing vapor rate below the flooding point, where fractionation efficiency would suffer

Control of column bottom level & effluent flow

- The column bottoms stream may be permitted to flow out of the column by gravity - there is no need to control this stream, unless it needs to be pumped to get to its destination, or unless it needs to be held back to prevent pressurized vapor from blowing through, as in a steam trap
- Most commonly, the bottoms feed stream is indeed pumped away, with a simple level control loop responding to load changes
- Frequently, the column bottoms stream is controlled by level cascaded to flow control, which achieves a more steady flow rate, important when the bottoms stream will pass through heat recovery & cooling equipment
- The bottoms level controller can also be cascaded to the reboiler steam controller, but this is generally only useful in columns where the bottoms product is a vapor side draw, or any other column with very small or intermittent need for bottoms flow - such as columns in which most of the feed will be vaporized, leaving only a small solids-containing bottom purge stream

Control of side draw streams, liquid & vapor

- Most often, liquid side draw streams are simply flow controlled, with the purity as measured by analytical techniques or by process temperature used by the human operator to adjust the flow set point
- Sometimes the liquid side draw flow responds well to cascade control from one of the column temperature controllers, by the same principle as distillate purity control
- Vapor side draws rarely are suitable for direct flow control in the vapor state, but if they are to be injected as a vapor into other equipment, this must be done, using an DP-type, turbine, or vortex shedding flow meter
- If vapor side draws are being condensed, it is usually easier & more reliable to measure the average flow rate of the condensed liquid leaving the vapor side draw condenser
- If the vapor side draw condenser liquid effluent is flow controlled & the condenser is a type which will tolerate partial flooding with liquid, then controlling the flow of liquid out will also achieve control of the vapor going into the condenser
- The vapor side draw flow rate setpoint may be taken from cascaded output from the column bottoms level or one of the column temperature controllers

Column pressure control - positive pressure systems

- In distillation systems which are meant to operate at atmospheric pressure, the system usually can have an open vent (often through a flame arrestor) to the air, or to a vapor collection header whose pressure is controlled near atmospheric by the end-of-pipe emission control device (oxidizer, condenser, scrubber or flare)

- If the gases in the vapor collection header should not be permitted to enter the distillation system, as they would at shut down, then a vent breathing control is required, usually using inert gas for breathing in
- In distillation systems that must operate at positive pressures, a pressure control loop using the column head pressure or condenser vent pressure is usually connected to a split range output, with one end of the range opening a vent valve & the other opening a pressurized gas source valve
- Control of condenser coolant temperature can also be used on one side of the split range pressure control, instead of a pressurizing gas source, but the dynamics of this control scheme make it slow to respond to load changes, so better for use in relative steady operations

Column pressure control - vacuum systems

- In distillation systems meant to operate at negative gauge pressure, usually an absolute pressure transmitter in the column head or in the condenser vent line is used in a simple pressure control loop, with output going to restrict a valve located between the distillation system & the vacuum source
- If the vacuum source is dedicated to this distillation system, the pressure control loop output alternatively may be used to open a discharge to feed recycle valve for a vacuum pump which operates as a fixed volume device, or an air bleed valve, or a steam supply valve for a steam ejector
- Rarely is gas makeup required for control of vacuum operations, because most real equipment experiences some air in-leakage, & most distillation feed materials contain some dissolved gas

Control of column pressure in multiple effect columns

- When one column's vapor is used to heat another column's reboiler, the first column's pressure depends on the condensing capability provided by the second column
- Generally, multiple effect trains of distillation columns experience pressure float, with increasing load increasing all the columns' pressures, & none except the lowest pressure column experiencing independently controllable fixed pressure
- The operating pressure of the upstream partner in a pair of multiple effect columns can be increased from its natural value dictated by heat transfer, only by restricting vapor flow (rarely practical) or by limiting condensing heat transfer surface available (usually practical) by permitting the level of condensate to rise in the downstream column's reboiler
- This control scheme causes inventory changes in the upstream column's condensate, & thus only is advisable for distillation systems on long term steady operating campaigns

How to create a new P&ID

- Start with equipment & main process pipe lines from the block flow diagram
- Add more detailed process piping
- Add utility supply & return piping
- Add instruments to measure & control the process & utilities
- Add valves for control, protection, or isolation
- Add pressure safety equipment for any section containing a vessel that can be blocked in
- Create & design control loops & interlocks

Start with a logical arrangement of equipment from the block flow diagram

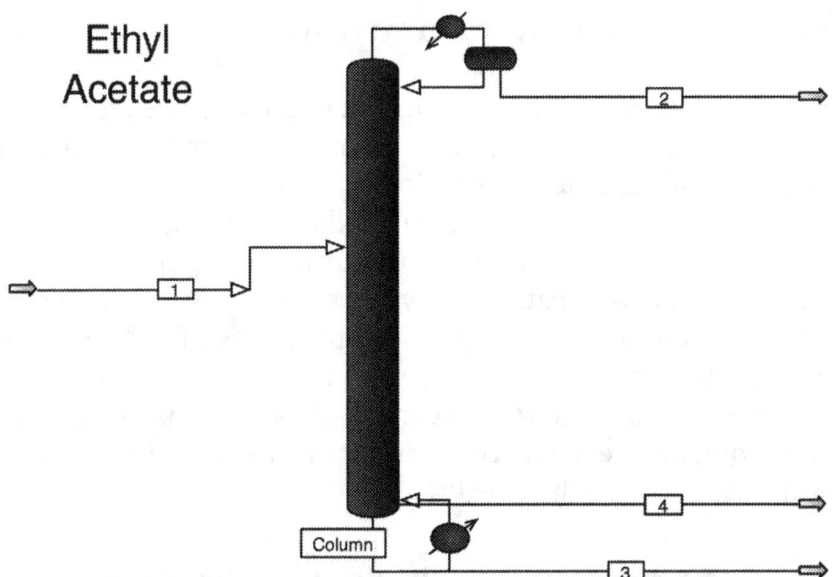

Add main process pipe lines from the block flow diagram

Add more detailed process path piping

Add utility supply & return & vent piping

Add measurement instruments for flow, pressure, temperature, level

Review the pipe line map to make certain that the equipment can be filled, emptied, & recycled

Add valves for control, protection, & isolation

Add safety equipment:

- Pressure relief PSV or PSE for any section containing a vessel that can be blocked in
- High level switches & low level switches
- High pressure switches if appropriate

Control strategy must be determined for:

- Feed input
- Heat input
- Column product purity control
- Column bottom effluent
- Reflux
- Column top distillate
- Other feed, product, or internal streams

Control loops must be created & designed to carry out selected control strategy

Add process indicators & safety interlocks

- Indicators added for information or data recording purposes
- Interlock devices included for protection of pumps & overall system safety

Continue development of P&ID after creation of all pipe lines, instruments, & controls:

- Instruments tagged & numbered for lists & specifications
- Every section of pipe tagged with size - fluid description - material code - insulation

Conduct P&ID operability review

Make certain that the system allows for:
- Filling
- Starting up
- Sampling
- Shutting down
- Emptying

Conduct P&ID environmental review:

- No uncontrolled vent emissions
- Seals on pumps & instruments up to good practices
- Means included to stop accidental spills or vent releases
- Plugs or blind flanges to prevent accidental releases
- Containment requirements for spills in the area

Conduct preliminary P&ID safety review

- Use HAZOP procedure or other safety review procedure
- Identify possible abnormal performance
- Include mitigating measures:
  - Check valves, interlocks, level switches, pressure safety valves, burst disks, extra valves or caps on drains & vent connections

Revise P&ID in accordance with recommendations of reviews:

- Operability
- Environmental
- Safety

Back-check the revised P&ID to make certain that:

- Review comments have been incorporated correctly
- No new problems have been created

Develop instrument list

- Instrument list will be extracted from the P&ID
- Can be created automatically or by hand
- Automation can be better if working from scratch, but needs non-interfering humans to accept its automatically generated numbering

Instrument list contents (for each instrument)

- Quantity
- Tag #
- Type
- Service
- Range
- Connections
- Material

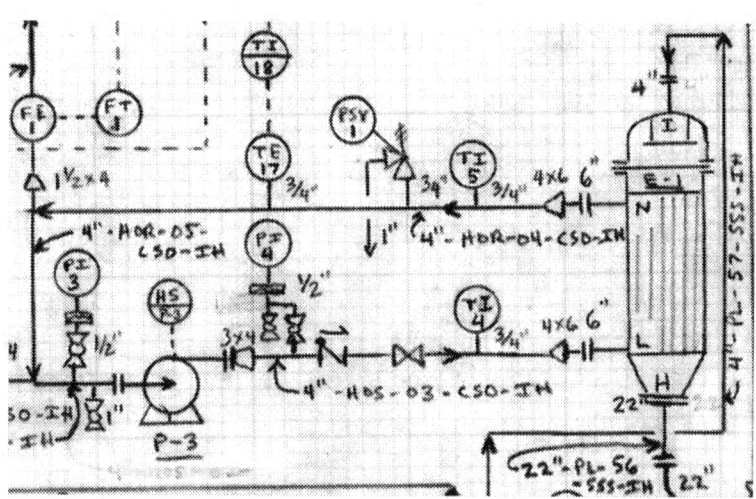

Instrument list categories

- Part 1: Field instruments, electrical & mechanical
- Part 2: Control panel devices, electrical & pneumatic
- Part 3: Control valves & actuated valves

Qty	Tag #	Type	Service	Range	Connections	Material
	DPT 9	2 Port DP	Column Vapor	0-25 mmHg	1/2" NPT	316
	FE 1	Orifice DP	Hot Oil	0-40 GPM	1 1/2" 150#	CS / SS trim
	FE 2	Coriolis Mass Flow w/Density E&H 63F	Terpene Distillate	0-9000 lb/hr	1" 150#	316 / 904L
	LSL 1	E&H M FTL 51	Citrus Oil	Fail Empty	1 1/2" 150#	316
	LSL 2	E&H M FTL 51	Terpenes	Fail Empty	1 1/2" 150#	316
	PI 1	PI / Dia. Seal	P-2 Suction	30"Vac-0-15	1/2" NPT	316
	PI 2	PI / Dia. Seal	P-2 Discharge	30"Vac-0-60	1/2" NPT	316
	PI 3	PI / Dia. Seal	P-1 Suction	0-100 psig	1/2" NPT	Brass
	PI 4	PI / Dia. Seal	P-1 Discharge	0-100 psig	1/2" NPT	Brass
	PI 5	Pressure Guage	Column Bot	30"Vac-0-15	1/2" NPT	316
	PI 7	Pressure Guage	Nitrogen	0-30 psig	1/2" NPT	316
	PI 8	Torr Pnl Guage	T-2 Abs Press	0-30 torr abs	1/2" NPT	316
	PSE 1	Burst Disk	Full Vac	Set 50 psig	2" 150#	316
	PSE 2	Burst Disk	Full Vac	Set 50 psig	2" 150#	316
	PSE 3	Burst Disk	Full Vac	Set 50 psig	6" 150#	316
	PSV 1	Liq Expansion	Hot Oil	Set 150 psig	3/4"x1" NPT	CS / SS trim
	PSV 2	Liq Expansion	Chilled Glycol	Set 100 psig	3/4"x1" NPT	CS / SS trim
	PSV 3	Liq Expansion	Chilled Glycol	Set 100 psig	3/4"x1" NPT	CS / SS trim
	PT 10	Absolute Press	Cond Vent	0-30 torr abs	1/2" NPT	316
	TE 10	RTD 100 ohm	Still Pot Temp	0-300 F	1 1/2" 150#	316
	TE 11	RTD 100 ohm	Col Bot Temp	0-300 F	1 1/2" 150#	316
	TE 12	RTD 100 ohm	Col Mid Temp	0-300 F	1 1/2" 150#	316
	TE 13	RTD 100 ohm	Col Top Temp	0-300 F	1 1/2" 150#	316
	TE 14	RTD 100 ohm	Col Vent T	0-300 F	3/4" NPT	316
	TE 16	RTD 100 ohm	Hot Oil Supply	50-450 F	3/4" NPT	Brass
	TE 17	RTD 100 ohm	Hot Oil Return	50-450 F	3/4" NPT	Brass
	TI 1	Dial Thermometer	Liquid Out of C-1 to P-2		3/4" NPT	316
	TI 15	Dial Thermometer	Chilled Glycol		3/4" NPT	Brass
	TI 2	Dial Thermometer	Chilled Glycol		3/4" NPT	Brass
	TI 3	Dial Thermometer	Chilled Glycol		3/4" NPT	Brass
	TI 4	Dial Thermometer	Hot Oil		3/4" NPT	Brass
	TI 5	Dial Thermometer	Hot Oil		3/4" NPT	Brass
	TI 6	Dial Thermometer	Chilled Glycol		3/4" NPT	Brass
	TI 7	Dial Thermometer	Cooled Folded Citrus Oil		3/4" NPT	316
	TI 8	Dial Thermometer	Chilled Glycol		3/4" NPT	Brass
	TI 9	Dial Thermometer	Chilled Glycol		3/4" NPT	Brass

Part 1: Field Instruments, Electric & Mechanical

| Part 2: Control Panel Devices, Electrical & Pneumatic | | | | | |
Qty	Tag #'s	Type	Service	Range	Connec-tions	Mate-rial
	DPIC 9	DP Controller	Column DP	0-25 mmHG		
	FIC 1	Flow Controller	Hot Oil Flow	0-40 GPM		
	FIC 2	Flow Controller	Terpene Distillate Flow	0-9000 lb/hr		
	HS 1	Switch 115V	Citrus Oil Feed	Open-Close		
	HS 2	Switch 115V	Nitrogen to C1	Open-Close		
	HS 3	Switch 115V	Nitrogen to T2	Open-Close		
	HS 4	Switch 115V	T1 Liq to T2	Open-Close		
	HS 5	Switch 115V	T2 Vent to T1	Open-Close		
	HS 6	Switch 115V	T2 to Evac Sys	Open-Close		
	HS P-1	Motor Switch	Feed Pump	Run-Stop		
	HS P-2	Motor Switch	Reboiler Circ	Run-Stop		
	HS P-3	Motor Switch	Hot Oil Circ	Run-Stop		
	HS P-4	Motor Switch	Terpenes Out	Run-Stop		
	I 1	P2 Interlock	LSL1 Stops P2			
	I 2	P4 Interlock	LSL2 Stops P4			
	PIC 10	Pres Controller	Condenser Vent Vacuum	0-30 torr abs		
	QIC 1	Heat Controller	Btu Input to Re-boiler	0-1M Btu/hr		
	TI 10	T Indicator	Still Pot	0-300 F		
	TI 11	T Indicator	Column Bot	0-300 F		
	TI 13	T Indicator	Column Top	0-300 F		
	TI 14	T Indicator	Column Vent	0-300 F		
	TI 16	T Indicator	Hot Oil Supply	50-450 F		
	TI 17	T Indicator	Hot Oil Return	50-450 F		
	TIC 12	T Controller	Column Mid	0-300 F		

Part 3: Control Valves & Actuated Valves					
Qty	Tag #	Type	Service	Range	Connec-tions	Mate-rial
	FV 1	Globe Control	Hot Oil 130-450 F	0-40 GPM	1 1/2" 150#	CS/SS Trim
	FV 2	V-Ball Control	Terpene Grav. Flow	0-22 GPM	1" 150#	316
	FY 1	I/P positioner	3-15 psig output	4-20 ma	NPT	Brass
	FY 2	I/P positioner	3-15 psig output	4-20 ma	NPT	Brass
	HV 1	Ball Vlv On-Off	Citrus Oil Feed	0-15 psig	2" SW	316
	HV 2	Ball Vlv On-Off	Nitrogen	0-15 psig	2" SW	316
	HV 3	Ball Vlv On-Off	Nitrogen	0-15 psig	2" SW	316
	HV 4	Ball Vlv On-Off	T-1 Liquid to T-2	0-15 psig	2" SW	316
	HV 5	Ball Vlv On-Off	T-2 Vent to T-1	0-15 psig	2" SW	316
	HV 6	Ball Vlv On-Off	T-2 to Evac System	0-15 psig	2" SW	316
	HY 1	Solenoid 3-port	Energize to Open	115 VAC	NPT	Brass
	HY 2	Solenoid 3-port	Energize to Open	115 VAC	NPT	Brass
	HY 3	Solenoid 3-port	Energize to Open	115 VAC	NPT	Brass
	HY 4	Solenoid 3-port	Energize to Open	115 VAC	NPT	Brass
	HY 5	Solenoid 3-port	Energize to Open	115 VAC	NPT	Brass
	HY 6	Solenoid 3-port	Energize to Open	115 VAC	NPT	Brass
	PV 10	V-Ball Control	Process Vacuum		4" 150#	316
	PY 10	I/P positioner	3-15 psig output	4-20 ma	NPT	Brass

PLC & DCS system software development

- Software development for PLC & DCS systems typically begins with URS development, review & approval
- Well-designed PLC & DCS systems look & feel the same to the plant operator, & can have all the same functionality
- PLC & DCS software design diverges at the FDS & DDS levels
- PLC control software typically is based on ladder logic, with human interface software developed separately, using an interface program (SCADA) which employs different tag numbers & names than the PLC operating ladder logic
- DCS control software usually is based on an object linking & embedding graphical control module concept & sequential function charts, with only one data base of tag numbers & names used both in control logic & in operator interface screens

GAMP control hardware & software life cycle

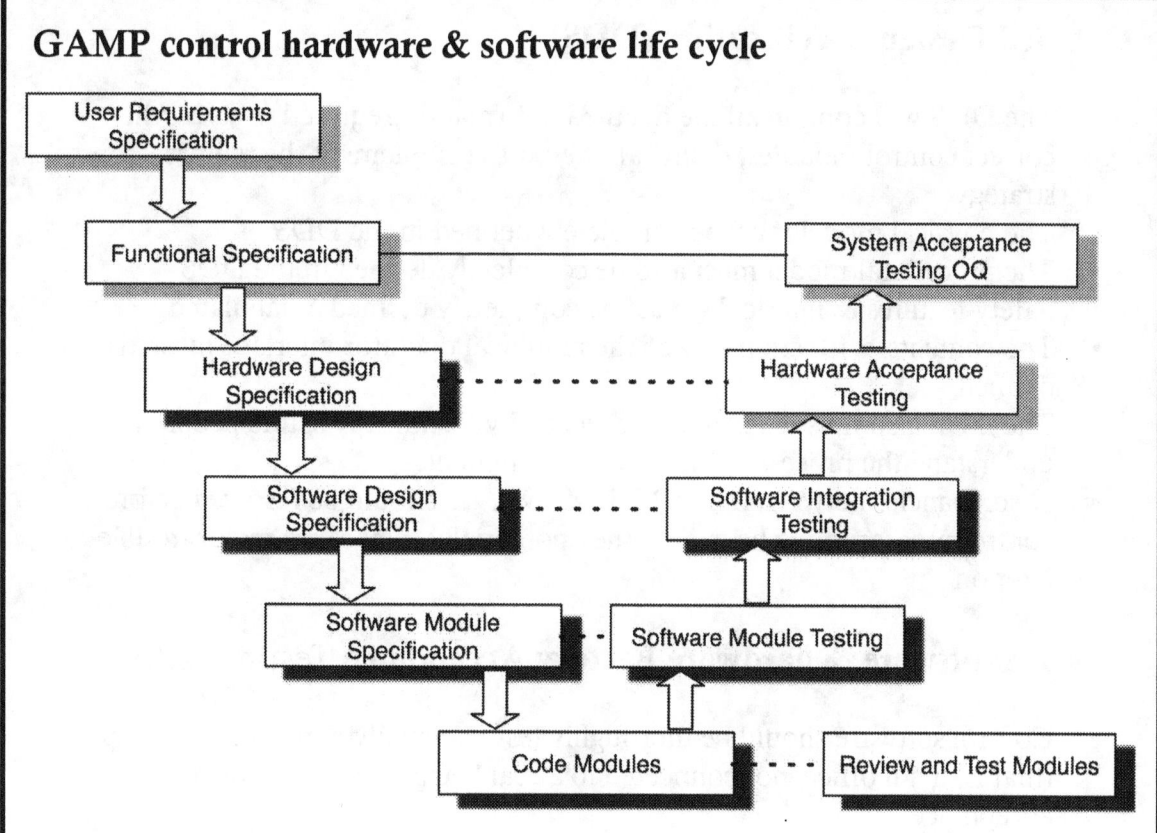

User Requirements Specification (URS)

- URS is the key document for control hardware specification & control software development
- The URS should be prepared jointly by a process engineer, a control engineer, & someone who knows the production methods, data collection requirements, & site safety standards

Functional Design Specification (FDS)

- The FDS is usually prepared by control engineers who will perform the control system programming
- The purpose of the FDS for a relatively small & simple system is to completely specify how the required control strategy (depicted on the P&ID's & in the URS) will be implemented on the selected control platform
- The purpose of the FDS for a relatively large or complex control system is to specify in broad terms what the approach will be to implementing the required control strategy
- The FDS should be read carefully by the team who prepared the URS, & discrepancies should be challenged now - not kept quiet, not challenged later!

Detailed Design Specification (DDS)

- The DDS will contain all the detailed information required to select the correct control modules, define all the I/O, & implement the required control strategy
- The physical model must be completely defined in the DDS
- The procedural model must also be completely defined in the DDS
- Safety features & interlocks must be completely defined & tabulated
- The contents & functionality of the graphical operator interface must also be defined
- The DDS also must be carefully checked by technically astute people who understand the process & the user's requirements
- Discrepancies between the DDS, FDS, & URS should be resolved before control programming begins, or they potentially may cause major trouble later on

Control software & hardware Factory Acceptance Testing (FAT)

- Control software should be thoroughly tested, usually with simulated inputs (that is, in an office, not connected to a real live process), before customer acceptance
- FAT testing should begin with the lowest level of control module integrations (I/O blocks) & proceed through to the highest level (sequential function charts or recipe execution)
- Hardware FAT testing should be performed using the completed & FAT tested software configuration, loaded on the hardware which will be included in the final installation - not substitutes!

Control software & hardware Site Acceptance Testing (SAT)

- Control software should be thoroughly tested after installation in the completed site hardware installation, within minimal simulated inputs & most I/O communicating with the field hardware of the actual process, before customer final acceptance of the site control installation
- SAT testing should begin with the lowest level of control module integrations (I/O blocks) & proceed through to the highest level (sequential function charts or recipe execution)
- Software to software interactions which have been thoroughly tested in the office FAT environment do not have be be completely tested again at the SAT, but some randomly sampled FAT tests should be repeated during the SAT to weed out possible problems
- Hardware FAT testing should be performed using the completed hardware installation with all wiring in place, & power supplied by the intended normal sources - no temporary wiring allowed!

Simplicity of design pays off in simplicity of testing

- Software which is designed with more complexity than is required will also need much more testing effort than is required
- Software testing is supposed to follow all possible software branches, so building too many branch points in & out of a software path (compared to programming several long, non-branching, parallel paths that perform the same function) multiplies the testing time & paper weight
- Simplicity of design pays off in simplicity of testing
- Think of this while preparing the URS & especially while reviewing the FDS

ISA-S88.01-1995 Batch control models & terminology

Standard for batch manufacturing control
- Defines models, concepts & terminology
- Promotes modularity & flexibility
- Emphasizes good practices for design & operation
- Applicable to batch distillation processes
- Can be used to manage a continuous distillation process operation from initializing parameters through startup, operation, & shutdown

S88 standard phases of control project execution

- Functional specification or FDS phase
  - Defines the automation requirements
- Design specification or DDS phase
  - Defines how the application software meets the functional specification requirements
- Implementation or configuration phase
  - Configuration of the application software
- Testing Phase
  - Verifies that the application software meets the functional specification & design specification requirements

GAMP control hardware & software life cycle

S88 model types: physical, procedural, process models

- Physical model
  - Defines the hierarchy of equipment used in the batch process & their physical points of interaction with the control system, I/O, discrete logic, & analog control loops
- Procedural model
  - Defines the control sequences & actions that enable equipment to perform the process steps
- Process model
  - Defines the process functionality from a high level

S88 Physical model

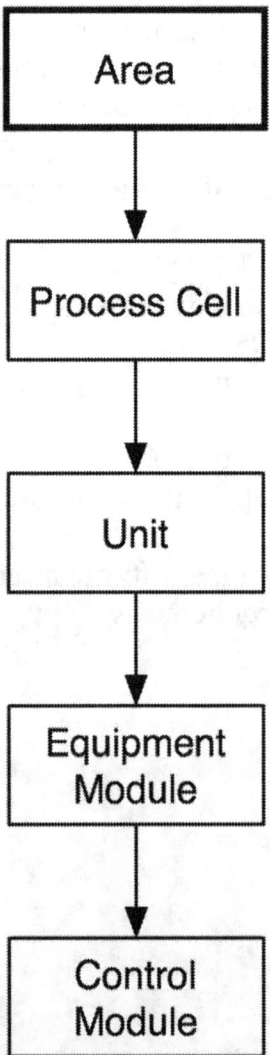

S88 Procedural model

```
┌──────────────┐
│  Procedure   │
└──────────────┘
       │
       ▼
┌──────────────┐
│     Unit     │
│  Procedure   │
└──────────────┘
       │
       ▼
┌──────────────┐
│  Operation   │
└──────────────┘
       │
       ▼
┌──────────────┐
│    Phase     │
└──────────────┘
```

S88 Process model

```
┌──────────────┐
│   Process    │
└──────────────┘
       │
       ▼
┌──────────────┐
│   Process    │
│    Stage     │
└──────────────┘
       │
       ▼
┌──────────────┐
│   Process    │
│  Operation   │
└──────────────┘
       │
       ▼
┌──────────────┐
│   Process    │
│   Action     │
└──────────────┘
```

S88 Procedural state matrix

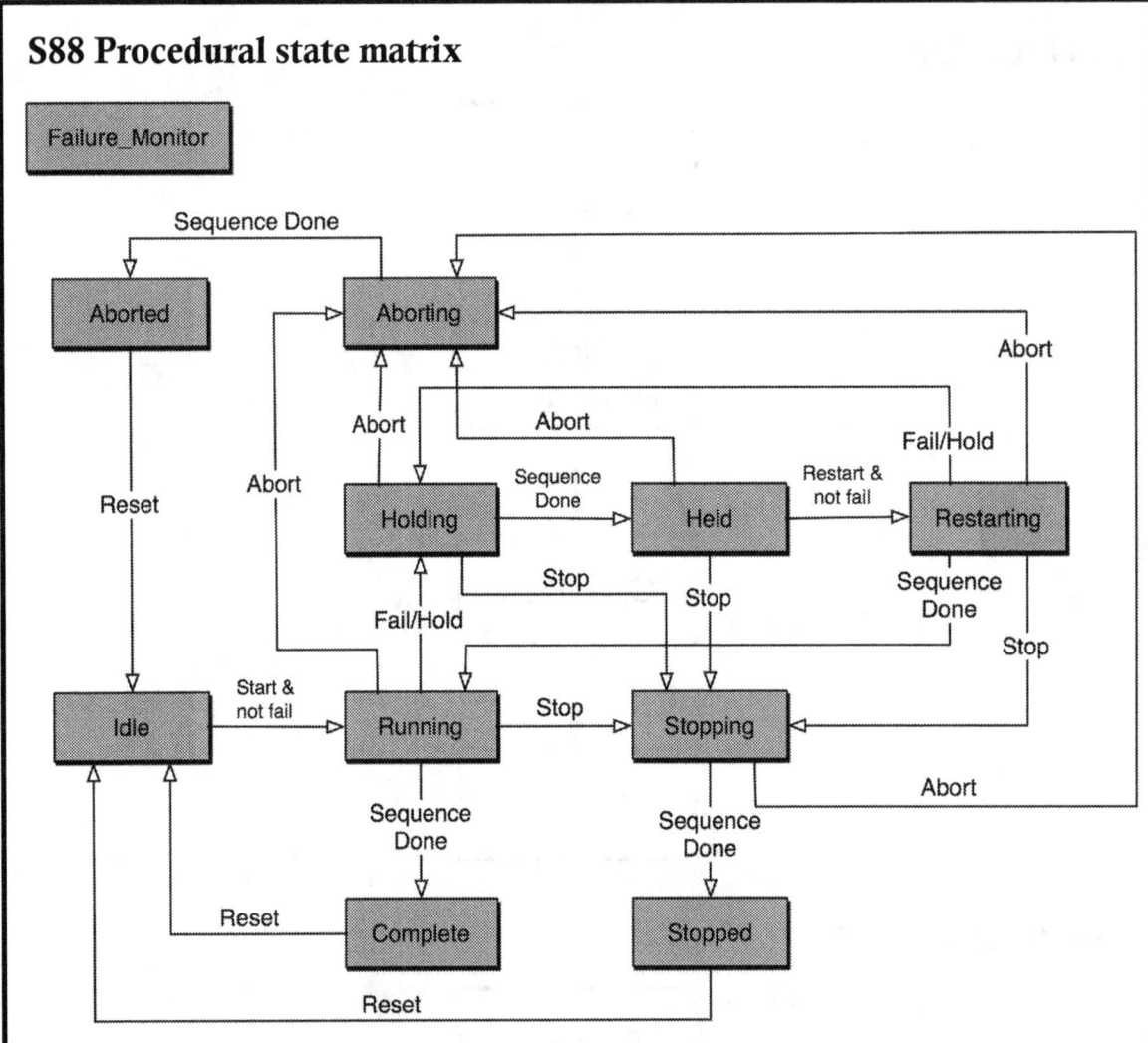

S88 Procedural state matrix

- **Idle** - No action. Waiting for a start command from the operator or a batch recipe instruction.
- **Running** - Normal operation of equipment. Begins when the start command is received.
- **Complete** - The running state has completed. Waits for a reset command to transition to idle.
- **Pausing** - (not pictured) The operator has issued a pause command during the running state. The running logic progresses to the next defined pause point.
- **Paused** - (not pictured) The running state is paused & no action is taken. At this point, the operator or a batch recipe can issue a hold, stop or abort command. When the operator issues the resume command, the running logic continues from the pause point & the state returns to running.

S88 Procedural state matrix

- **Holding** - Equipment is placed in a safe state. The running state is disrupted & placed in holding when an exception to normal operation is detected or the operator issues the hold command.
- **Held** - Holding state has completed. No actions are taken. At this point, the operator or a batch recipe can issue a restart, stop or abort command.
- **Restarting** - The restart command has been issued by the operator while the state is held. Takes action to return equipment to normal operation. Once restarting finishes, it transitions to the running state.

S88 Procedural state matrix

- **Stopping** - Equipment is placed in a safe state. The current state is disrupted when the operator issues the stop command.
- **Stopped** - Stopping state has completed. At this point, the recipe cannot be restarted.
- **Aborting** - Equipment is placed in a safe state. The current state is disrupted when the operator issues the abort command.
- **Aborted** - Aborting state has completed. At this point, the recipe cannot be restarted.

Pressure control equipment module summary

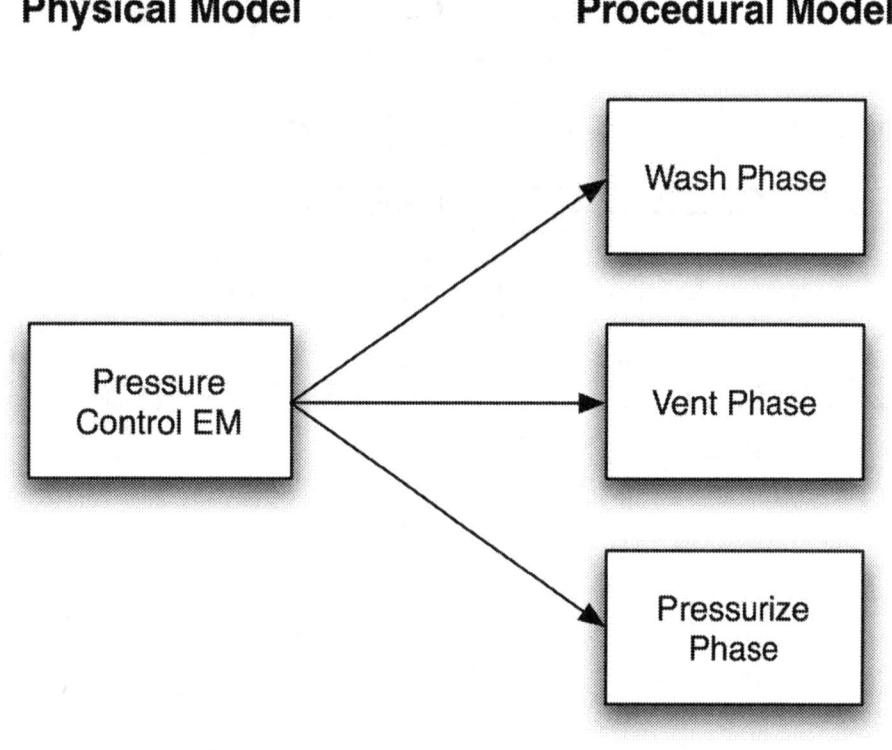

Elements such as flow control loops are in the physical model

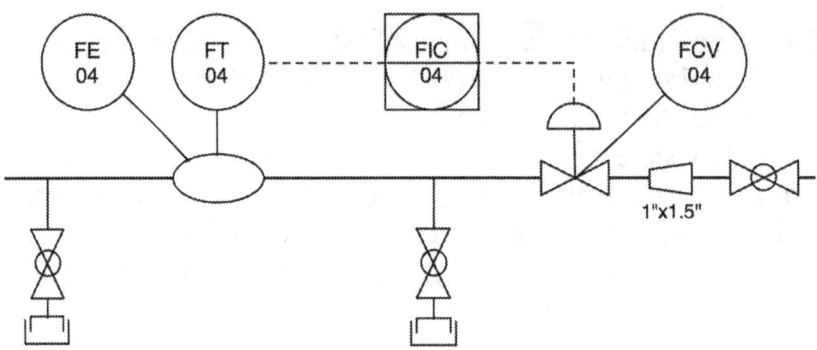

Sequential function charts subdivided into phases & steps make up the procedural model

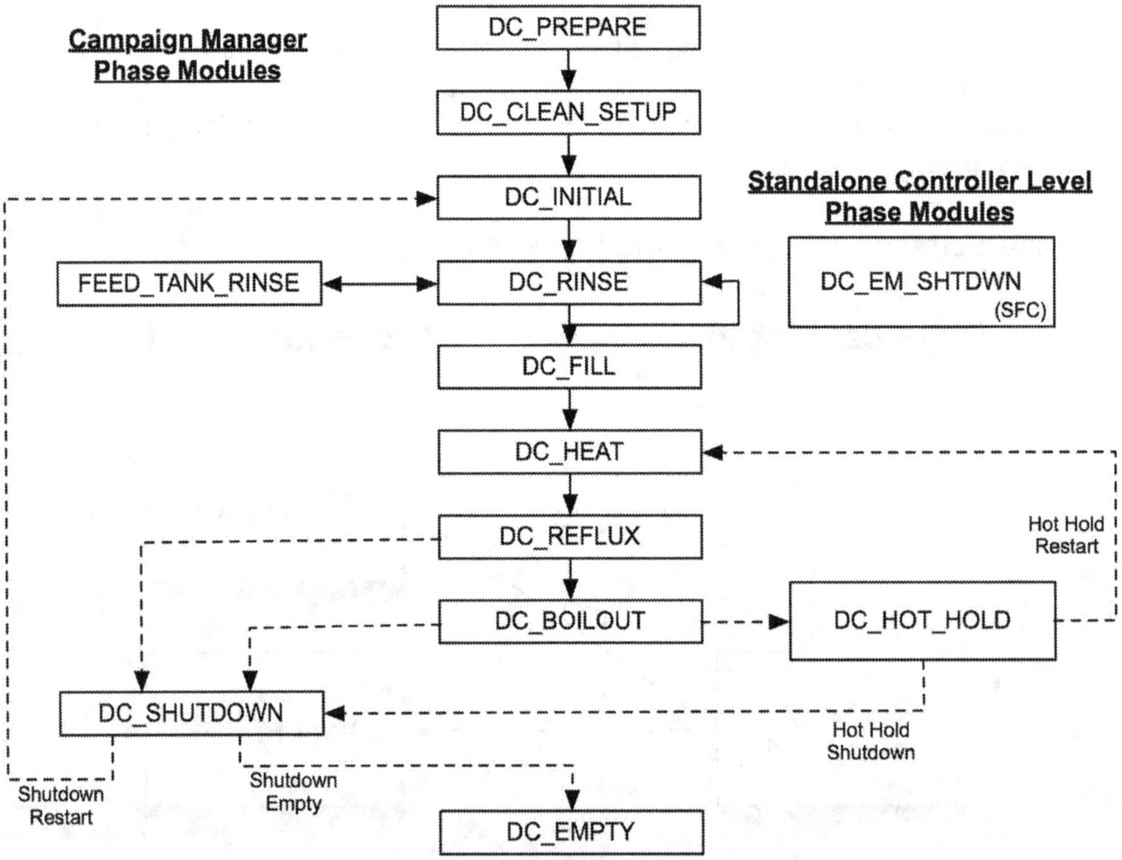

S88 batch recipe procedural model used to control a distillation system

- The following presents a summary description of the implementation of a distillation production operation recipe & a system cleaning operation recipe on a DCS control platform running a batch executive following the S88 procedural model
- This information is intended to define the flow of the procedural model during these operations, & explain how the phases are implemented within each recipe
- The next three images show a 3-D model of the distillation system being controlled by the recipes defined here
- Recipe descriptions are presented in the next sections, with a phase map for each operation & a brief summary of the phase actions, followed by safety interlock charts

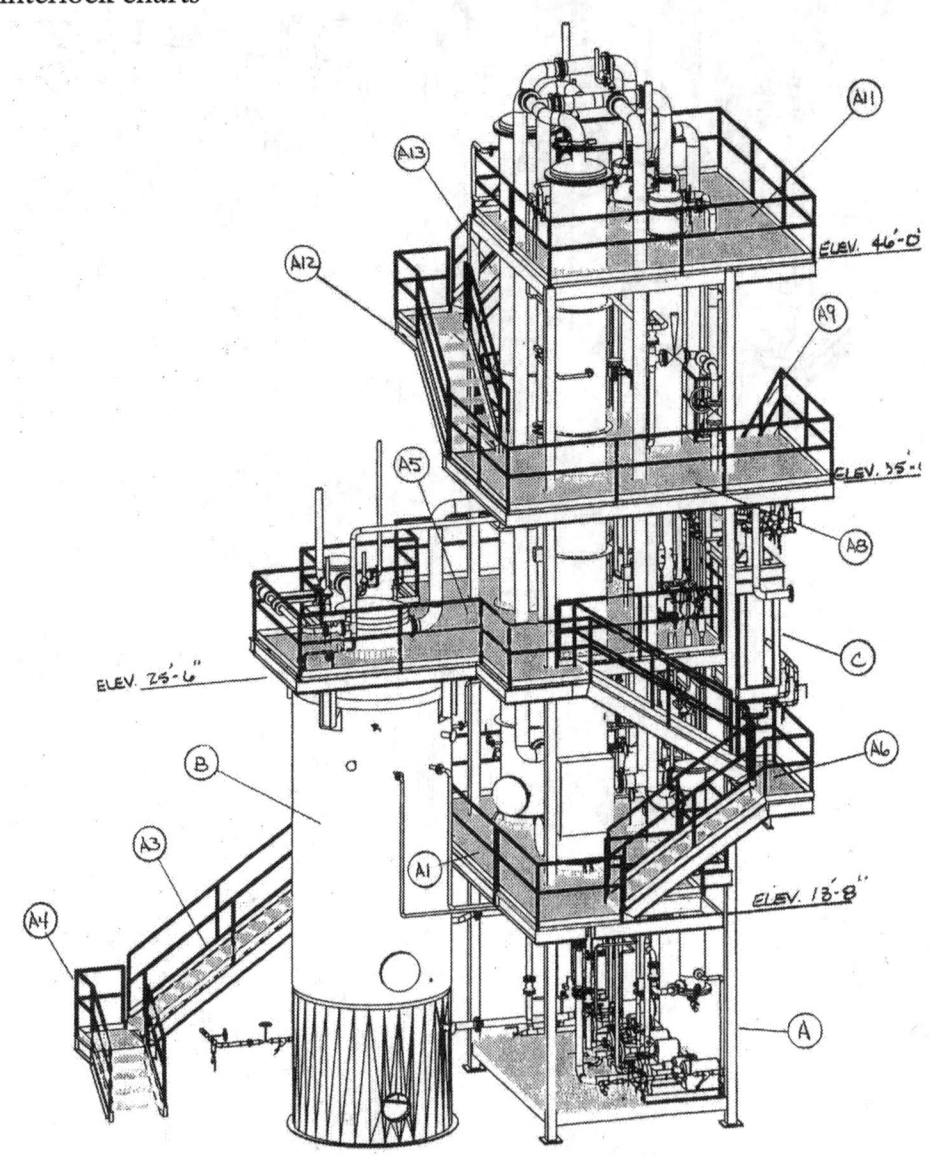

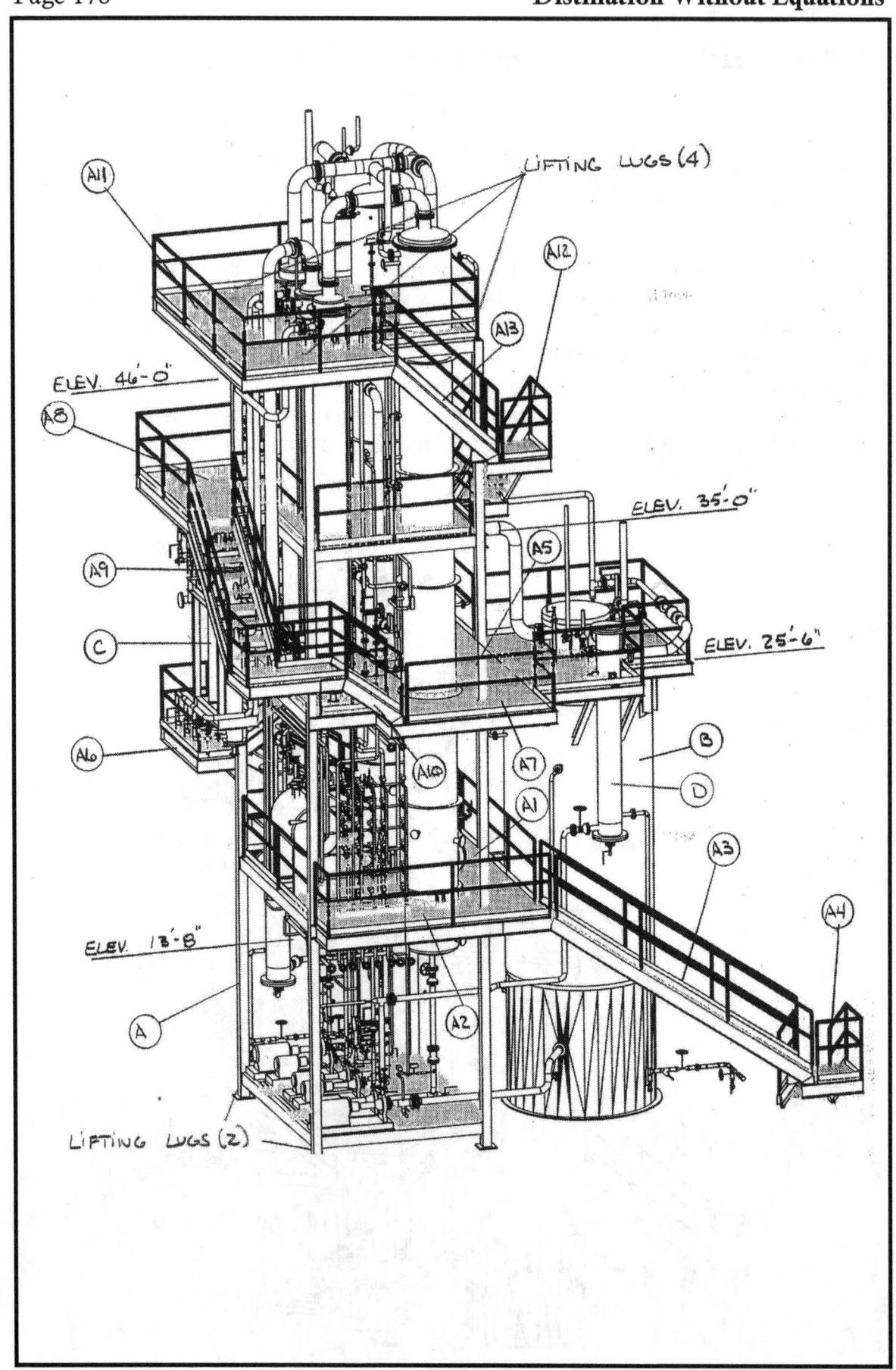

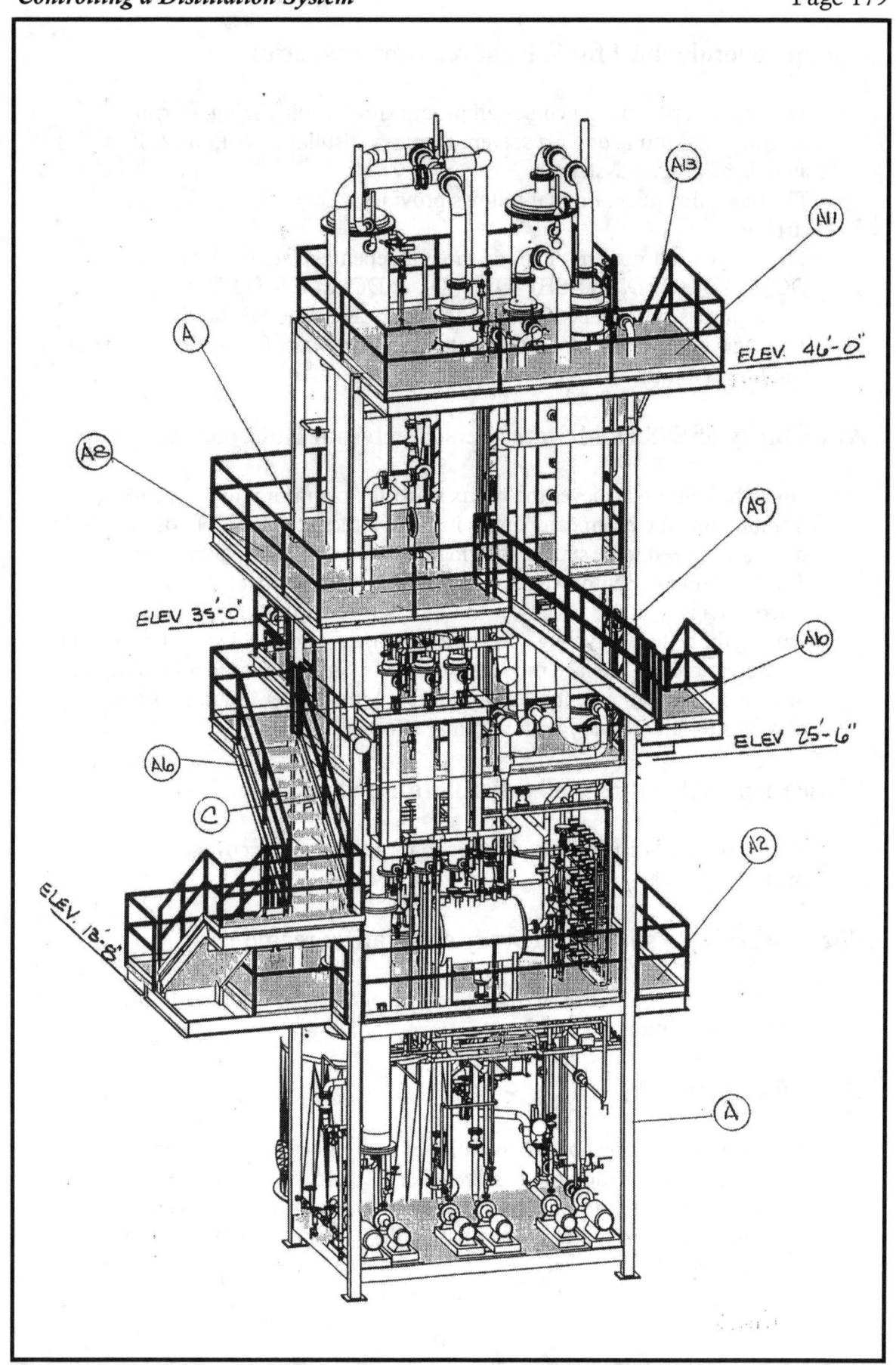

S88 procedural model for solvent recovery operation

- The procedural model includes all procedures for initializing, starting, running, & shutting down a solvent recovery distillation column & its associated process systems
- The block flow diagram that follows provides an overview of the phase structure
- The recipe itself does not include three independent stand alone phases, DC_HOT_HOLD, DC_SHUTDOWN, & DC_EM_SHTDWN
- These phases run independently at the controller's module level
- The logic in these phases executes at all times & waits for process or operator conditions to initiate its logic

Availability of S88 Hold States versus external Hold phases

- Once the column achieves the reflux state, the operator will be prohibited from holding the batch campaign - instead, either the Hot Hold or Shutdown phases will need to be started to bring the system to a safe holding state
- The Hot Hold & Shutdown phases interface with the Batch phases via discrete flags
- During the reflux, recycling, & recovery phases, a hot hold condition executes the hot hold logic to place the system process equipment into a holding state, in which the column continues distillation with a lower input, thus allowing it to return to full recovery with minimal effort

Phase map - solvent recovery distillation recipe

- The next image displays the phase map for the solvent recovery recipe operation

Phase overview - solvent recovery distillation recipe

- The following are brief descriptions of the phases shown above for the solvent recovery operation

DC_PREPARE

- Checks to see that no control modules or equipment modules associated with the unit have simulation mode selected or interlocks bypassed
- If any modules are in simulation mode or have interlocks bypassed the operator will be prompted & the phase will not continue until the problem has been addressed

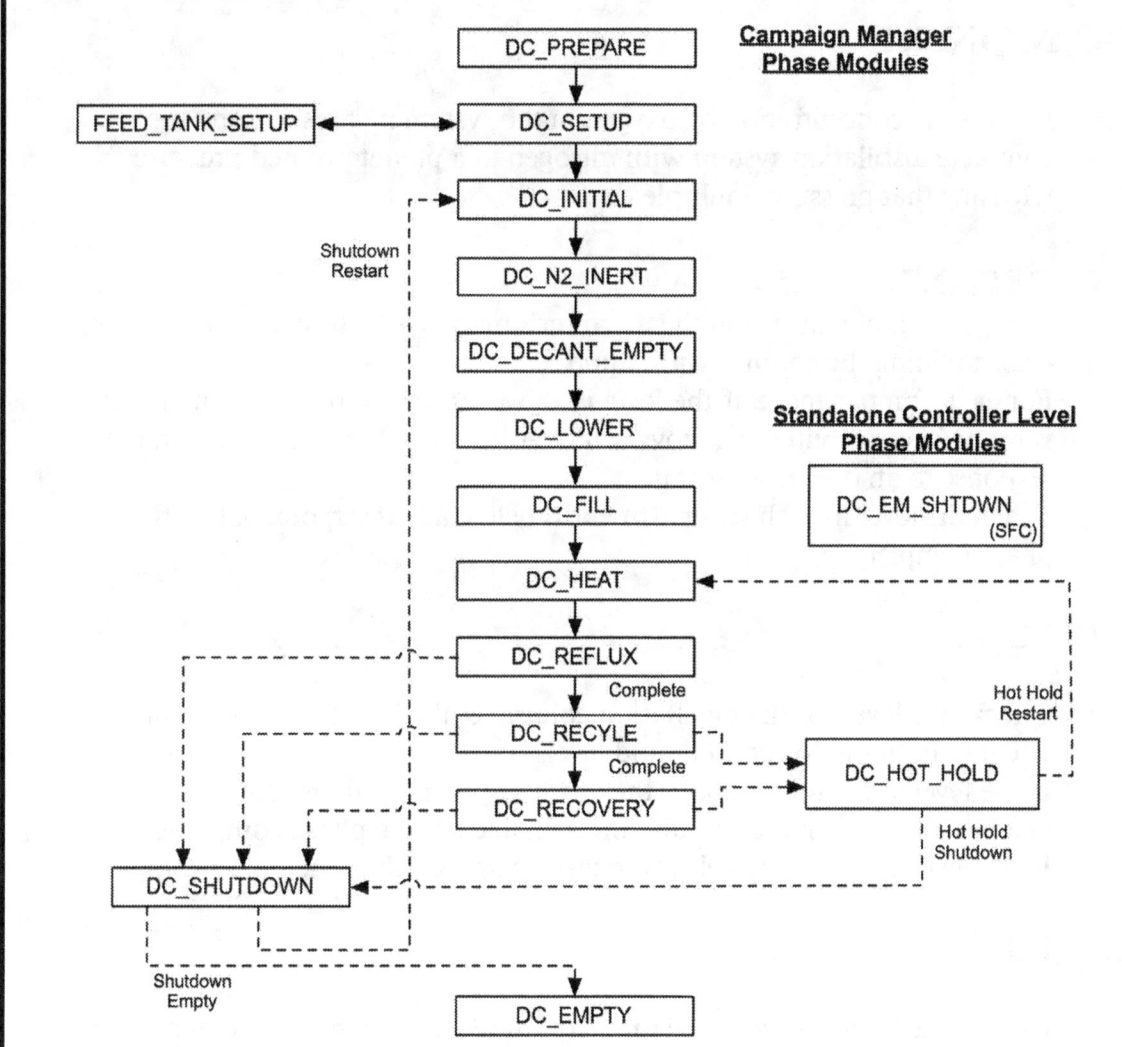

DC_SETUP

- Configures the column control strategies for the correct setup based on recipe
- The phase sets unit parameters, configures master PID controllers for the cascaded loops &, where multiple process variables are used, configures the recipe selected PV

DC_INITIAL

- Runs the DC_SET_CASCADE composite, which sets equipment to Cascade mode, & drives it to its fail-safe state
- All automated solenoid valves are set to the fail-safe condition (usually closed), controllers are set to manual with 0% output, & all pumps are stopped
- The phase also runs the DC_LOCAL_DISABLE composite, which disables operator control of automated equipment

DC_N2_INERT

- Lowers the concentration of oxygen in the system by pressurizing the complete distillation system with nitrogen to a predetermined pressure & releasing that pressure, multiple times

DC_DECANT_EMPTY

- Pumps out any material in the decanter's main & overflow compartments prior to filling the column with liquid
- For each compartment, if the level is above a given setpoint, the material is pumped out to either a recovery tank, or a waste location based on the response to an operator prompt
- When the level in each compartment is below a low setpoint value, then the phase completes

DC_LOWER

- Checks the levels of the distillation column & the batch pot & determines whether or not to lower the liquid levels of the unit
- If the level in the unit is above the given setpoint, liquid is removed until it reaches a target value, recirculation is enabled & the phase completes
- If the level in the unit is below the given setpoint, then the phase completes

DC_FILL

- Checks the levels of the distillation column & the batch pot & determines whether or not to raise the level to a setpoint value within a given dead band
- If the level in the unit is below the setpoint, liquid will be fed until it reaches a target value
- Once the level in the unit is within the specified setpoint range, recirculation is enabled & the phase completes
- If the level in the unit is above the given setpoint at the start of the phase, recirculation is enabled & the phase completes

DC_HEAT

- Initiates the heating sequence for the distillation unit
- The steam valves (distillation column & batch pot as necessary) will be ramped open at a specified rate, to a desired steam flow rate value
- The phase continues until the level of material in the decanter increases by 5%, & then the phase completes

DC_REFLUX

- Configures the column equipment for achieving total reflux
- Depending on the recipe setting of decanter mode, miscible, immiscible heavy, or immiscible light, the phase controls the reflux from the correct decanter compartment level controller
- The controller output is cascaded to the setpoint of the reflux controller

DC_RECYCLE

- Transitions the column from a total reflux condition to the recycle to feed configuration in which all distillate, product, & waste streams are recycled back to the feed tank
- During this phase all column flow, level & temperature controls are functioning in automatic
- This phase enables the automatic column feed control, the decanter distillate take-off, the decanter waste source take-off, the Column & Batch Pot bottoms waste, & the side draw product flow

DC_HOT_HOLD

- Transfers the process equipment to a holding state, in which the column continues distillation with a lower input, thus allowing it to return to full recovery with minimal effort

DC_SHUTDOWN

- Performs a controlled shutdown on the Distillation Column Unit equipment

DC_EMPTY

- Empties the distillation column & blows nitrogen through the column & associated equipment after the column as been shutdown

DC_EM_SHTDWN

- Performs a rapid emergency shutdown of the Distillation Column Unit equipment
- This logic is not executed in a phase
- Instead, this logic resides in sequential function chart embedded in a controller resident control module
- The logic drives all equipment to their safe states, & can either be operator initiated from a button on the operator interface, or will be initiated from the control module on detection of emergency shutdown conditions

Input parameters for solvent recovery recipe operation

Parameter	Description	Settings
recSolvFeedRate	Solvent Feed Setpoint	Type: Real Lo Rng: 0.0 Hi Rng: 9000 Default: 4000 Units: lbs/hr
recSolvFeedVlv	Sovent Feed Valve	Type: Integer Named Set Value = 0: T1 Tray 21 (FV27) Value = 1: T1 Tray 30 (FV28) Value = 2: T2 Tray 37 (FV29) Value = 3: T2 Tray 49 (FV30) Value = 4: Batch Pot (FV20) Default: FV27
recWaterEnabled	Water Feed Enabled	Type: Discrete Named Set Value = 0: Disabled Value = 1: Enabled Default: Disabled
recWaterFeedRate	Water Feed Setpoint	Type: Real Lo Rng: 0.0 Hi Rng: 9000 Default: 0 Units: lbs/hr
recWaterFeedVlv	Water Feed Valve	Type: Integer Named Set Value = 0: Tray 30 (FV25 - Column T1) Value = 1: Tray 41 (FV26 - Column T2) Default: FV25
recBatchPot	Batch Pot Enabled	Type: Discrete Named Set Value = 0: Disabled Value = 1: Enabled Default: Disabled
recColStmFlowSP	Steam Flow Setpoint to Column Reboiler E-2	Type: Real Lo Rng: 0.0 Hi Rng: 6000 Default: 2400 Units: lbs/hr

Parameter	Description	Settings
recBPSStmFlowSP	Steam Flow Setpoint to Batch Pot Reboiler E-1	Type: Real Lo Rng: 0.0 Hi Rng: 6000 Default: 2400 Units: lbs/hr
recRatioHeatVap	Ratio of Solvent to Water Heat of Vaporization	Type: Real Lo Rng: 0.000 Hi Rng: 2.000 Default: 0.35 Units: N/A
recProductType	Column Product Type	Type: Integer Named Set Value = 1: Top Value = 2: Liquid Side Draw Value = 3: Vapor Side Draw Default: Top
recDecantMode	Decanter Mode of operation	Type: Integer Named Set Value = 1: Miscible Value = 2: Immiscible Heavy Value = 3: Immiscible Light Default: Miscible
recPriApprovalTnk	Primary approval tank used for first time selection of product destination	Type: Discrete Named Set Value = 0: 30-D-130 Value = 1: 30-D-135 Default: 30-D-130
recDistMstrTemp	Master column temperature sensor used for distillate control	Type: Integer Named Set Value = 1: T1 Bottom (TT_51902) Value = 2: Tray 07 (TT_51903) Value = 3: Tray 13 (TT_51904) Value = 4: Tray 28 (TT_51905) Value = 5: Tray 30 (TT_51906 Value = 6: T2 Bottom (TT_51907) Value = 7: Tray 36 (TT_51908) Value = 8: Tray 48 (TT_51909) Value = 9: Tray 56 (TT_51910) Value = 10: Tray 60 (TT_51911) Value = 11: None Default: None

Parameter	Description	Settings
recDistTempSP	Master column temperature setpoint for distillate control	Type: Real Lo Rng: 30 Hi Rng: 130 Default: None Units: Deg C
recLSDMstrTemp	Master column temperature sensor for liquid side draw product flow control. This value cannot equal the value set for recDistMstrTemp.	Type: Integer Named Set Value = 1: T1 Bottom (TT_51902) Value = 2: Tray 07 (TT_51903) Value = 3: Tray 13 (TT_51904) Value = 4: Tray 28 (TT_51905) Value = 5: Tray 30 (TT_51906 Value = 6: T2 Bottom (TT_51907) Value = 7: Tray 36 (TT_51908) Value = 8: Tray 48 (TT_51909) Value = 9: Tray 56 (TT_51910) Value = 10: Tray 60 (TT_51911) Value = 11: None Default: None
recLSDTempSP	Master column temperature setpoint for liquid side draw product flow control	Type: Real Lo Rng: 30 Hi Rng: 130 Default: None Units: Deg C
recSLDOutletVlv	Liquid Side Draw Outlet Valve	Type: Integer Named Set Value = 0: None Value = 1: Tray 48 (FV31) Value = 2: Tray 52 (FV32) Value = 3: Tray 56 (FV33) Default: None

S88 procedural model for cleaning/decontamination operation

- The Cleaning/Decontamination recipe includes all procedures necessary to rinse, fill, heat & boil out the distillation column for cleaning purposes
- The diagram that follows provides an overview of the phase structure
- As in the previous section, the recipe itself does not include three independent stand alone phases, DC_HOT_HOLD, DC_SHUTDOWN, & DC_EM_ SHTDWN
- These phases run independently at the module level of the controller
- The logic for these phases executes at the time process conditions or the operator initiate the logic

Availability of S88 Hold States versus use of external Hold Phases

- Once the column achieves the reflux state, the operator will be prohibited from holding the batch campaign - instead, either the Hot Hold or Shutdown phase will need to be run to bring the system to a safe holding condition
- The Hot Hold & Shutdown phases interface with the Batch phases via discrete flags
- During the Reflux & Boilout phases, a hot hold condition executes the hot hold logic to place the system process equipment to a holding state, in which the column continues distillation with a lower energy input, thus allowing it to return to full recovery with minimal effort
- Once the Hot Hold state is achieved, the operator is prompted to either shutdown or return to column recovery
- Once the Shutdown state is achieved, the operator is prompted to either continue to the rinse & purge phase or return & restart the cleaning procedure

Phase map - distillation system cleaning/decontamination recipe

- The next image displays the phase map for the system cleaning/ decontamination recipe operation

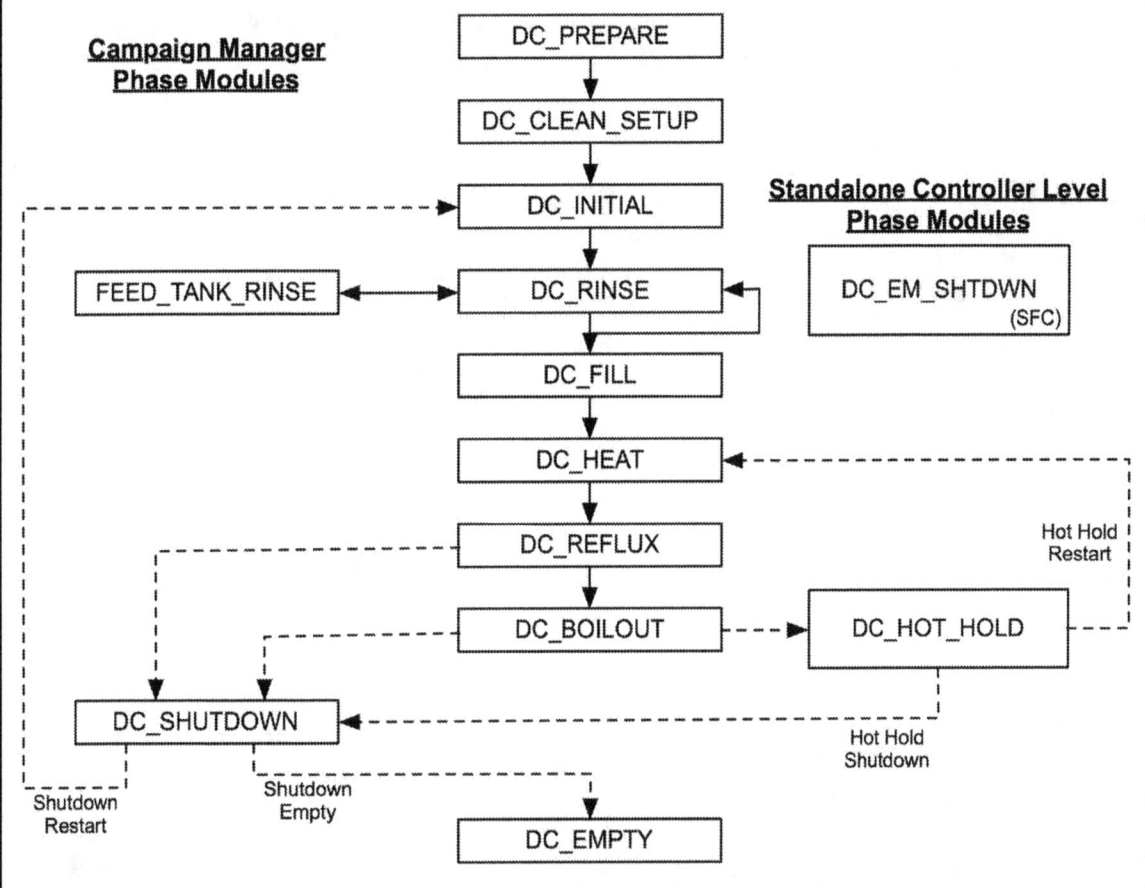

Phase overview - system cleaning/decontamination recipe

- The following are brief descriptions of the phases shown above which are unique to the cleaning/ decontamination sequence, not included among the modular phases which cleaning/decontamination shares in common with the distillation operation

DC_RINSE

- Fills the Distillation Column Unit equipment with process water, circulates the water for a specified amount of time, & discharges the contents of the column via the bottom discharge route until the column is empty

DC_BOILOUT

- Performs a boil out of the Distillation Column Unit equipment including all receiver vessels
- It flushes condensate through all the various outlet lines, to purge residue from past recovery operations
- The phase is intended to execute during the cleaning recipe after the column has been filled, heated to boiling, & achieved total reflux

Input parameters for cleaning recipe operation

Parameter	Description	Settings
recBPStmFlowSP	Steam Flow Setpoint to Batch Pot Reboiler E-1	Type: Real Lo Rng: 0.0 Hi Rng: 6000 Default: 4000 Units: lbs/hr
recColStmFlowSP	Steam Flow Setpoint to Column Reboiler E-2	Type: Real Lo Rng: 0.0 Hi Rng: 6000 Default: 4000 Units: lbs/hr
recDistMstrTemp	Master column temperature sensor used for distillate control	Type: Integer Named Set Value = 1: T1 Bottom (TT_51902) Value = 2: Tray 07 (TT_51903) Value = 3: Tray 13 (TT_51904) Value = 4: Tray 28 (TT_51905) Value = 5: Tray 30 (TT_51906 Value = 6: T2 Bottom (TT_51907) Value = 7: Tray 36 (TT_51908) Value = 8: Tray 48 (TT_51909) Value = 9: Tray 56 (TT_51910) Value = 10: Tray 60 (TT_51911) Value = 11: None Default: None
recDistTempSP	Master column temperature setpoint for distillate control	Type: Real Lo Rng: 30 Hi Rng: 130 Default: None Units: Deg C
recFdTnkLevel	Feed Tank Level Setpoint	Type: Real Lo Rng: 0 Hi Rng: 80 Default: 10 Units: %

Parameter	Description	Settings
recFdTnkRecirc	Feed Tank Recirculation Time Setpoint	Type: Real Lo Rng: 0.0 Hi Rng: 120.0 Default: 10.0 Units: min
recBatchPot	Batch Pot Enabled	Type: Discrete Named Set Value = 0: Disabled Value = 1: Enabled Default: Disabled
recBPRecirc	Batch Pot Recirculation Time Setpoint	Type: Real Lo Rng: 0.0 Hi Rng: 120.0 Default: 10.0 Units: min
recBPRinseCount	Batch Pot Number of Rinses	Type: Integer Lo Rng: 0 Hi Rng: 10 Default 3 Units: N/A
recDistCol	Distillation Column Enabled	Type: Discrete Named Set Value = 0: Disabled Value = 1: Enabled Default: Disabled
recDCRecircTimer	Distillation Column Reciculation Time Setpoint	Type: Real Lo Rng: 0.0 Hi Rng: 120.0 Default: 10.0 Units: min
recDCRinseCount	Distillation Column Number of Rinses	Type: Integer Lo Rng: 0 Hi Rng: 10 Default 3 Units: N/A

Other procedural model logic always active & standing by

- E-Stop - Module Level Logic executed as a result of a predetermined triggering condition:
  - Any time E-Stop is triggered, the currently active phase transitions to the HELD state & then completes
- ShtDwn - Module Level Phase executed as a result of a triggering condition:
  - Any time ShtDwn is called for, the currently active phase transitions to the HELD state & then completes in a manner determined by a polling of discrete flags indicating the system condition existing before ShtDwn was triggered

ESTOP_MONITOR interlock actions have no phase dependence

Logic Module	No.	Tag	Description	Module Level Phase Executed	Alarm	Trip	Active Phases and Results
ESTOP_MONITOR	1	XS_FIRE	Fire Alarm System Contact	ES		Switch = 0	E-Stop At All Times - No Phase Dependence - All Phases Hold & Abort Batch
	2	XS_ESTOP	Distillation Area Estop Pushbutton	ES		Switch = 0	
	3	XS_PF	Power Failure	ES		Switch = 0	
	4	PT01	Batch Pot High Pressure	ES	PAHH	20 psia	
	5	PT01	Batch Pot Low Pressure	ES	PALL	13 psia	
	6	PT02	Tower T1 Bottoms High Pressure	ES	PAHH	20 psia	
	7	PT02	Tower T1 Bottoms Low Pressure	ES	PALL	13 psia	
	8	PT03	Tower T1 Top High Pressure	ES	PAHH	20 psia	
	9	PT03	Tower T1 Top Low Pressure	ES	PALL	13 psia	
	10	PT04	Tower T2 Bottoms High Pressure	ES	PAHH	20 psia	
	11	PT04	Tower T2 Bottoms Low Pressure	ES	PALL	13 psia	
	12	PT05	Tower T2 Top High Pressure	ES	PAHH	20 psia	
	13	PT05	Tower T2 Top Low Pressure	ES	PALL	13 psia	

SHUTDOWN_MONITOR actions are phase dependent

Logic Module No.	Tag	Description	Module Level Phase Executed	Time Delay	Alarm	Trip
		SHUTDOWN_MONITOR				
14	LSH01	Batc Pot high level	NS	15 sec	LSH	Switch = 0
15	LSH02	Tower T1 high level	NS	15 sec	LSH	Switch = 0
16	LSH03	Tower T2 high level	NS	15 sec	LSH	Switch = 0
17	LT01	Batch Pot low level	NS	15 sec	LALL	5%
18	LT02	Tower T1 low low level	NS	15 sec	LALL	5%
19	LT03	Tower T2 low low level	NS	15 sec	LALL	5%
20	LT07	Light Phase Decanter Compartment high level	NS	15 sec	LAH	95%
21	LT08	Heavy Phase Decanter Compartment high level	NS	15 sec	LAH	95%
22	PT02/PT03	Tower T1 high differential pressure	NS	30 sec	PDAHH	6 psi
23	PT04/PT05	Tower T2 high differential pressure	NS	30 sec	PDAHH	6psi
24	TT13	Batch Pot Bottoms Cool exit discharge high temp	NS		TAHH	60 °C
25	TT14	Tower Bottoms Cool exit discharge high temp	NS		TAHH	60 °C
26	TT16	Liquid Side Draw Condenser discharge high temp	NS		TAHH	60 °C
27	TT17	Main Condenser discharge high temperature	NS		TAHH	60 °C
28	LSLL	Feed Tank low low level	NS		LALL	Switch = 0

No.	Prepare	N2 Inert	Initial	Setup	Decant Empty	Lower	Fill	Heat	Reflux	Recycle	Recovery	Hot Hold	Shut down	Empty	Rinse	Boilout	FT Setup	FT Rinse
14									ShtDwn			ShtDwn				ShtDwn		
15								Batch Abort	ShtDwn			ShtDwn				ShtDwn		
16									ShtDwn			ShtDwn				ShtDwn		
17																		
18																		
19																		
20										ShtDwn	ShtDwn							
21										ShtDwn	ShtDwn							
22		Batch Hold																
23		Batch Hold						Batch Abort	ShtDwn			ShtDwn				ShtDwn		
24								Batch Abort	ShtDwn			ShtDwn				ShtDwn		
25								Batch Abort	ShtDwn			ShtDwn				ShtDwn		
26																		
27																		
28																		

PHASE_FAILURES logic module has phase dependent output

Logic Module No.	Tag	Description	Module Level Phase Executed	Trip
		PHASE_FAILURES		
29	FV01	Steam Valve to Batch Pot Reboiler interlock condition	Phase Specific Condition	Switch = 0
30	FV02	Steam Valve to Column Reboiler interlock condition		Switch = 0
31	XL01	Pump P1 Not Running		Switch = 0
32	IL1	Pump P1 Interlock Conditions		Switch = 0
33	XL02	Pump P2 Not Running		Switch = 0
34	ILP2	Pump P2 Interlock Conditions		Switch = 0
35	XL04	Pump P4 Not Running		Switch = 0
36	ILP4	Pump P4 Interlock Conditions		Switch = 0
37	XL05	Pump P5 Not Running		Switch = 0
38	ILP5	Pump P5 Interlock Conditions		Switch = 0
39	XL06	Pump P6 Not Running		Switch = 0
40	ILP6	Pump P6 Interlock Conditions		Switch = 0

No.	Prepare	N2 Inert	Initial	Setup	Decant Empty	Lower	Fill	Heat	Reflux	Recycle	Recovery	Hot Hold	Shut down	Empty	Rinse	Boilout	FT Setup	FT Rinse
29																		
30																		
31						Batch Hold*	Batch Hold*	Batch Hold						Batch Hold				
32																		
33														Batch Hold				
34									ShtDwn	ShtDwn	ShtDwn	ShtDwn						
35					Batch Hold									Batch Hold		ShtDwn		
36																		
37														Batch Hold				
38																		
39						Batch Hold								Batch Hold				
40																		

Individual parameter-based interlock actions are always active

No.	Tag	Description	Alarm	Trip	Equipment Trips
1	XS_FIRE	Fire Alarm System Contact		Switch = 0	All
2	XS_ESTOP	Distillation Area E-Stop Pushbutton		Switch = 0	All
3	XS_PF	Power Failure		Switch = 0	All
4	PT01	Batch Pot High Pressure	PAHH	20 psia	FV01/FV02 Steam
5	PT01	Batch Pot Low Pressure	PALL	13 psia	FV01/FV02 Steam
6	PT02	Tower T1 Bottoms High Pressure	PAHH	20 psia	FV01/FV02 Steam
7	PT02	Tower T1 Bottoms Low Pressure	PALL	13 psia	FV01/FV02 Steam
8	PT03	Tower T1 Top High Pressure	PAHH	20 psia	FV01/FV02 Steam
9	PT03	Tower T1 Top Low Pressure	PALL	13 psia	FV01/FV02 Steam
10	PT04	Tower T2 Bottoms High Pressure	PAHH	20 psia	FV01/FV02 Steam
11	PT04	Tower T2 Bottoms Low Pressure	PALL	13 psia	FV01/FV02 Steam
12	PT05	Tower T2 Top High Pressure	PAHH	20 psia	FV01/FV02 Steam
13	PT05	Tower T2 Top Low Pressure	PALL	13 psia	FV01/FV02 Steam

No.	Tag	Description	Time Delay	Alarm	Trip	Equipment Trips
14	LSH01	Batch Pot high level	15 sec	LSH	Switch = 0	FV01/FV02 Steam
15	LSH02	Tower T1 high level	15 sec	LSH	Switch = 0	FV01/FV02 Steam
16	LSH03	Tower T2 high level	15 sec	LSH	Switch = 0	FV01/FV02 Steam
17	TT06	Tower T1 high top temperature		TAHH	115° C	FV01/FV02 Steam
18	TT11	Tower T2 high top temperature		TAHH	115° C	FV01/FV02 Steam
19	XL01	Reboiler pump P1 stopped			Switch = 0	FV01 Steam
20	XL02	Reboiler pump P2 stopped			Switch = 0	FV02 Steam
21	FV03	Pump P1 Suction Valve FV03 closed			Switch = 0	Pump P1
22	FV08A, FV21	Pump P1 Discharge Valves FV08A & FV21 closed			Switch = 0	Pump P1
23	FV64	Nitrogen blow through valve FV64 opened			Switch = 1	Pump P1

No.	Tag	Description	Alarm	Trip	Equipment Trips
24	LSL01	Batch Pot low level switch	LSL	Switch = 0	Pump P1
25	FV04	Pump P2 Suction Valve FV04 closed		Switch = 0	Pump P2
26	FV08B, 22, 23	Pump P2 Discharge Valves FV08A, FV22 & FV23 closed		Switch = 0	Pump P2
27	FV65	Nitrogen blow through valve FV65 opened		Switch = 1	Pump P2
28	LSL02	Tower T1 low level switch	LSL	Switch = 0	Pump P2
29	FV13	Pump P3 Discharge Valves FV13 closed		Switch = 0	Pump P3
30	FV16	Pump P3 Suction Volve FV16 closed		Switch = 0	Pump P3
31	FV68	Nitrogen blow through valve FV68 opened		Switch = 1	Pump P3
32	LSL06	Vapor Side Draw Condenser low level	LSL	Switch = 0	Pump P3
33	FV06	Pump P4 Suction Valve FV06 closed		Switch = 0	Pump P4
34	FV14, FV61	Pump P4 Discharge Valves FV14 & FV61 closed		Switch = 0	Pump P4
35	FV69	Nitrogen blow through valve FV69 opened		Switch = 1	Pump P4

No.	Tag	Description	Alarm	Trip	Equipment Trips
36	LSL04	Heavy Phase Decanter Compartment low level switch	LSL	Switch = 0	Pump P4
37	FV07	Pump P5 Suction Valve FV07 closed		Switch = 0	Pump P5
38	FV59, FV62	Pump P5 Discharge Valves FV59 & FV62 closed		Switch = 0	Pump P5
39	FV70	Nitrogen blow through valve FV70 opened		Switch = 1	Pump P5
40	LSL05	Light Phase Decanter Compartment low level switch	LSL	Switch = 0	Pump P5
41	FV05	Pump P6 Suction Valve FV05 closed		Switch = 0	Pump P6
42	FV11, FV12	Pump P6 Discharge Valves FV11 & FV12 closed		Switch = 0	Pump P6
43	FV66	Nitrogen blow through valve FV66 opened		Switch = 1	Pump P6
44	LSL03	Tower T2 low level	LSL	Switch = 0	Pump P6

When interlocks are defined & programmed correctly, this never happens!

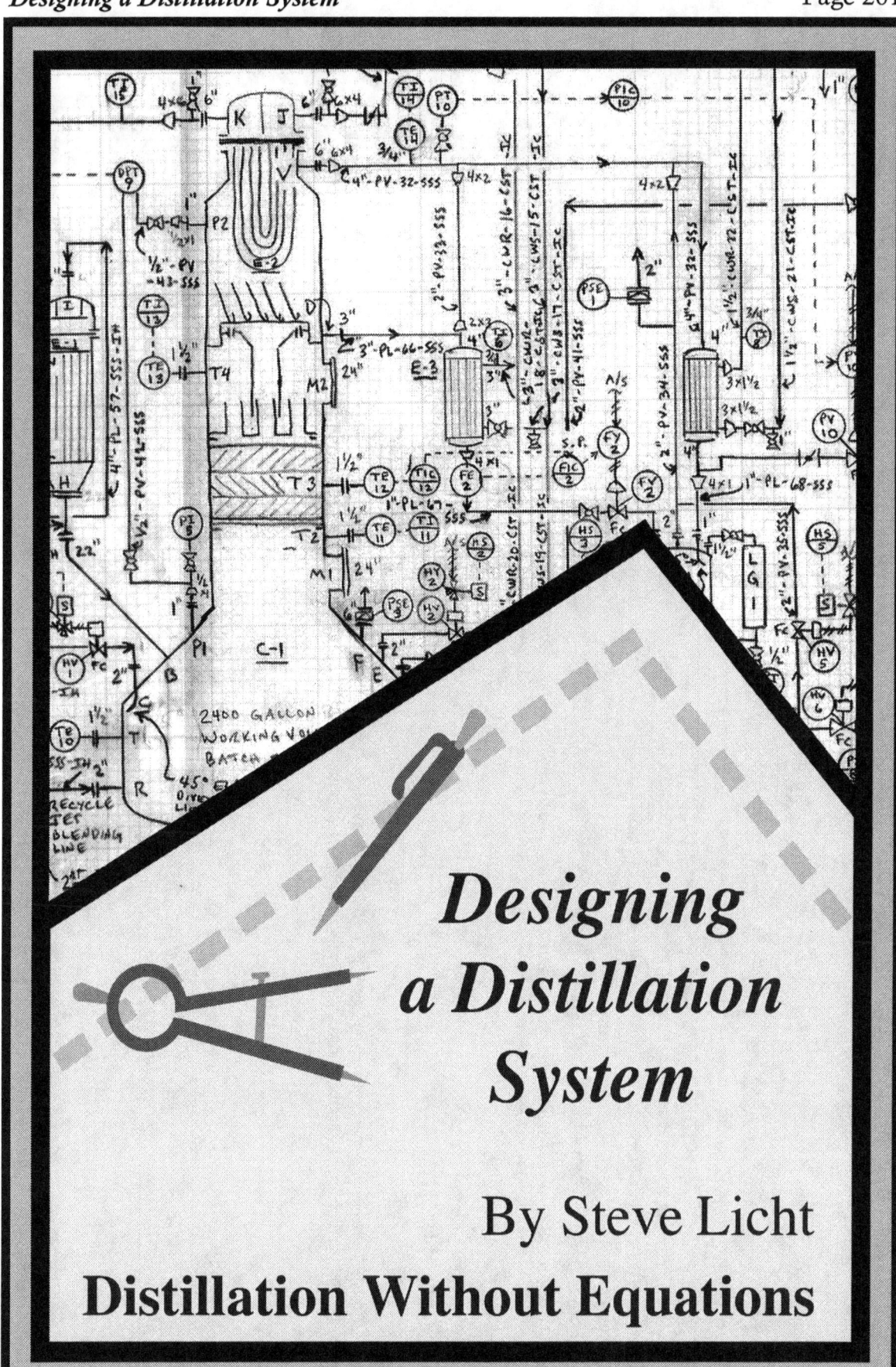

Designing a Distillation System

By Steve Licht

Distillation Without Equations

Designing a Distillation System

Part of the

Distillation Without Equations
Technical Seminar

By Steve Licht

Distillation system design

How to specify:
- Process duties
- Materials
- Equipment
- Instruments
- Utilities

How to conduct:
- Safety reviews
- Environmental reviews
- Operability reviews

Questions to be answered by the design basis

- What is coming in & going out?
- What is flowing inside of the system?
- What is heating?
- What is boiling?
- What is condensing?
- What is cooling?
- What is contacting?
- What is pumping?
- What requires control?
- What needs monitoring & recording?
- What needs sampling?
- What other things are standing by, just in case they are needed?

Design basis information

Collect materials & information that describe the new intended system & previous reference systems
- Block flow diagrams, material balances, preliminary P&ID's, reference system P&ID's, design memos, specification sheets, equipment quotes, catalog cuts, utility conditions

Categorize & sort
- Find the most current versions of design information
- Sort out what may be true or false -- Try not to repeat mistakes of the past
- Determine what is right or wrong about the preliminary design & the referenced prior designs

What is coming in & going out?

- This information should fit into a table showing quantities, compositions, temperatures, & pressures
- There may be only one set of data, or multiple sets for multipurpose systems
- Firm up this information before proceeding any further
- No holes in this data should be allowed

Design basis flow information: 1: Feed 2: Waste 3: Product

IPA/Water			
Stream ID	1	2	3
Temperature (C)	15.0	81.9	82.8
Mass Flow (KG/HR)	800.000	150.924	649.079
Mass Fraction			
Water	0.002	0.010	100 PPM
ISOPR-01	0.998	0.990	1.000
METHA-01			
DIBUT-01			

Design basis flow information: 1: Feed 2,4: Waste 3: Product

IPA/Methanol/Water				
Stream ID	1	2	3	4
Temperature (C)	15.0	66.3	82.8	74.6
Mass Flow (KG/HR)	500.000	27.000	446.000	27.000
Mass Fraction				
Water	0.002	0.008	102 PPM	0.028
ISOPR-01	0.948	0.249	1.000	0.790
METHA-01	0.050	0.744	918 PPB	0.182
DIBUT-01				

What is flowing inside of the system?

Ethyl
Acetate

Column

Ethyl Acetate system flow: 1: Feed 4: Product 2,3: Wastes

Stream Name	1	2	4	3
Stream Description	Feed	OVHDS Waste	Main Product	CRUD
Phase	Liquid	Liquid	Vapor	Liquid
KG/HR	2310.000	1751.617	548.383	10.000
Temperature C	10.000	64.721	82.587	82.910
Pressure BAR	4.000	1.000	1.209	1.220
Molecular Wight	69.989	65.710	87.950	88.086
Weight Comp. Percents				
H2O	4.3290	5.7033	0.0182	0.0007
Ethanol	6.2771	8.2564	0.0689	0.0093
EOAC	72.6407	63.9912	99.7706	99.9513
IPA	0.5195	0.6659	0.0611	0.0146
MEAC	16.2338	21.3832	0.0811	0.0241
Weight Comp. Rates KG/HR				
H2O	100.0000	99.9000	0.1000	0.0001
Ethanol	145.0000	144.6210	0.3780	0.0009
EOAC	1677.9999	1120.8801	547.1246	9.9951
IPA	12.0000	11.6632	0.3353	0.0015
MEAC	374.9999	374.5526	0.4450	0.0024
Enthalpy M*KCAL/HR	0.031	0.077	0.069	0.000

Ethyl Acetate distillation column liquid & vapor flows, all trays

Tray	Flow Rate	
	Liquid	Vapor
	KG-MOL/ HR	KG-MOL/ HR
1	179.2	
2	181.0	205.8
3	181.1	207.6
4	181.3	207.7
5	181.5	207.9
6	181.8	208.2
7	182.1	208.5
8	182.4	208.8
9	225.2	209.1
10	225.8	218.9
11	226.4	219.4
12	227.0	220.0
13	227.7	220.7
14	227.8	221.3
15	225.9	221.5
16	222.3	219.5
17	222.8	216.0
18	224.2	216.4
19	224.9	217.9
20	225.3	218.6
21		225.1

Ethyl Acetate column internal flows - Graphical representation

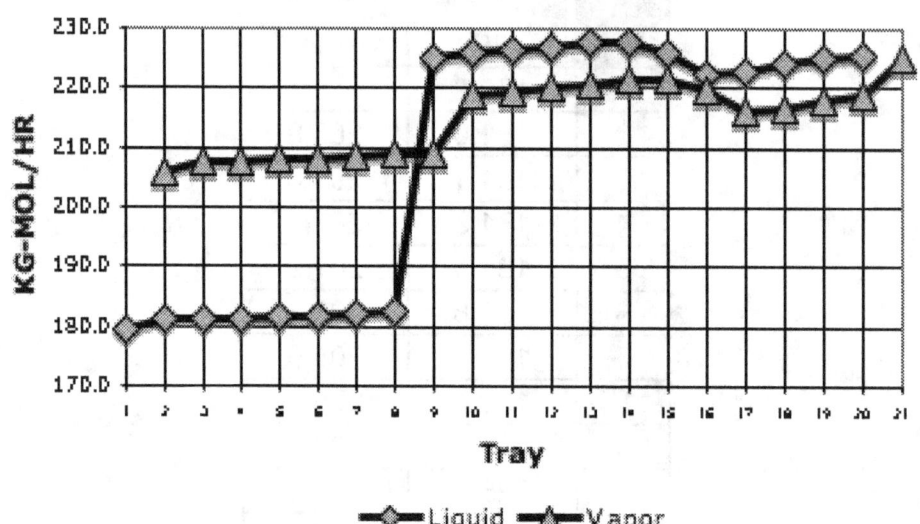

Rates Ethyl-Acetate

What is heating?

Ethyl Acetate column temperature profile

Tray	Temperature C
1	64.7
2	68.2
3	69.0
4	69.6
5	70.1
6	70.5
7	70.9
8	71.2
9	71.5
10	72.0
11	72.4
12	72.8
13	73.1
14	73.5
15	74.1
16	75.9
17	79.2
18	81.3
19	83.1
20	82.6
21	82.9

Ethyl Acetate column top to bottom temperature graph

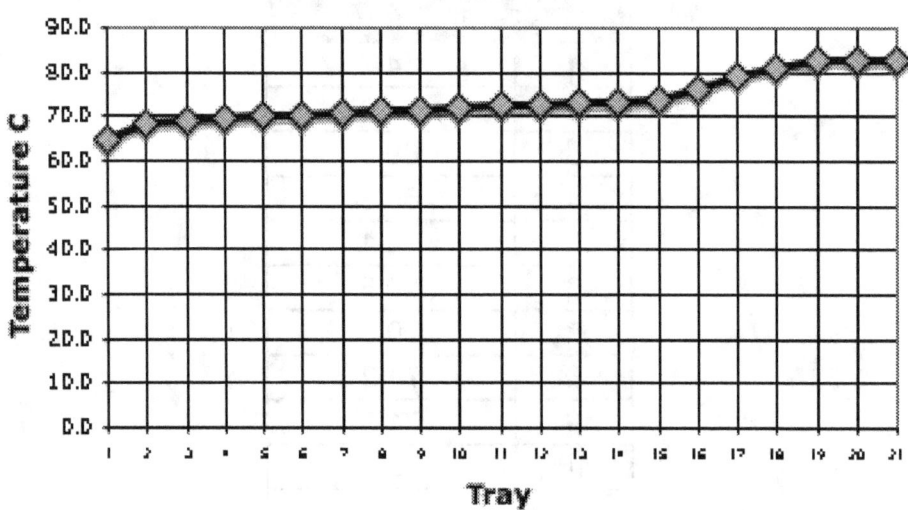

What is boiling?

Ethyl Acetate system flows: Mass & heat, in & out, at T & P

Column Summery							
Tray	Temperature C	Pressure BAR	Net Flow Rates				Duties
			Liquid	Vapor	Feed	Product	
			KG-MOL/HR				K*KCAL/ HR
1	64.7	1.00	179.2			26.7	-1.5920
2	68.2	1.02	181.0	205.8			
3	69.0	1.03	181.1	207.6			
4	69.6	1.04	181.3	207.7			
5	70.1	1.05	181.5	207.9			
6	70.5	1.06	181.8	208.2			
7	70.9	1.07	182.1	208.5			
8	71.2	1.08	182.4	208.8			
9	71.5	1.09	225.2	209.1	33.0		
10	72.0	1.10	225.8	218.9			
11	72.4	1.11	226.4	219.4			
12	72.8	1.13	227.0	220.0			
13	73.1	1.14	227.7	220.7			
14	73.5	1.15	227.8	221.3			
15	74.1	1.16	225.9	221.5			
16	75.9	1.17	222.3	219.5			
17	79.2	1.18	222.8	216.0			
18	81.3	1.19	224.2	216.4			
19	83.1	1.20	224.9	217.9			
20	82.6	1.21	225.3	218.6		6.2	
21	82.9	1.22		225.1		0.1	1.7076

What is condensing?

What is cooling?

What is contacting?

Ethyl Acetate column internal flows: Liquid & vapor by stage

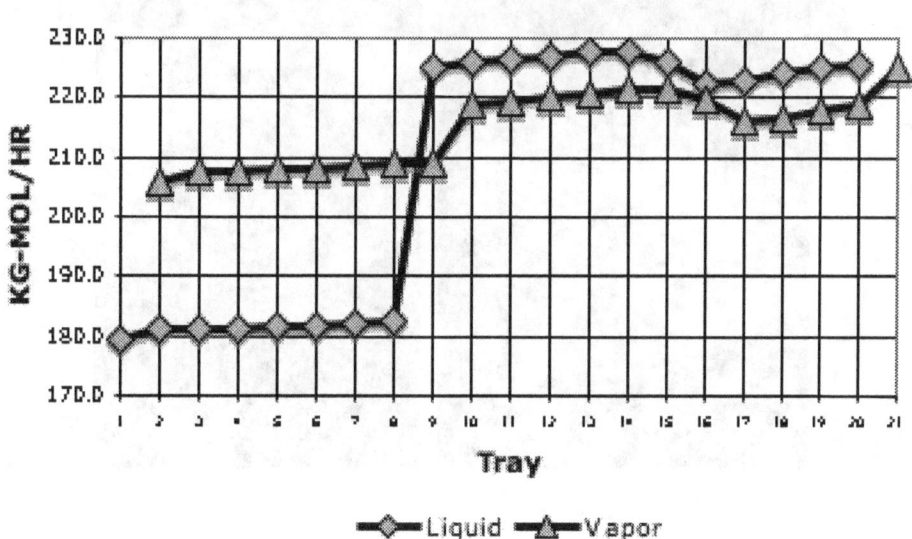

Rates Ethyl-Acetate

Underside of a valve tray with round valves & downcomers

One part of multi-section valve tray with rectangular valves

Trays are mounted horizontally inside distillation columns

What is pumping?

What requires control?

What needs monitoring & recording?

What needs sampling?

What other things are standing by, just in case they are needed?

Define Major Equipment List

- This list contains all of the equipment required to make the actions of the block flow diagram happen
- Equipment that will heat, boil, condense, cool, pump, contact, decant, or contain materials

Big Bang Begins

Specify equipment for the Major Equipment List

- Determine duties, materials, types, models
- What needs to be done by equipment, to accomplish what the block flow diagram describes?
- What equipment characteristics, capabilities, & capacities will be required?

Design equipment in order:

- Distillation columns & internals
- Reboilers & condensers
- Other heat exchangers
- Pumps
- Decanters
- Batch pot & product receivers
- Other tanks & agitators

Equipment specification sheets

- Heat exchangers (condenser, reboiler, coolers, others)
- Shell & tube exchangers are most common, but do not ignore other types:
  - Plate & frame
  - Spiral
  - Air-cooled finned

Plate & frame heat exchangers: Small passages & small size

Shell & tube heat exchanger baffles, rods, tubesheet, tubes

Shell & tube heat exchanger tube bundle lifted from shell

Shell & tube heat exchanger tubesheet with rolled-in tubes

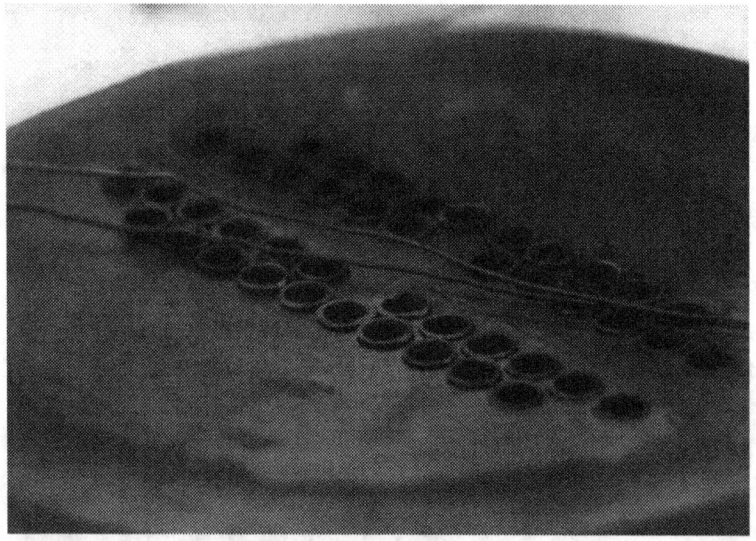

Shell & tube heat exchanger two-pass bonnet & gasket

Shell & tube heat exchanger specification sheets

- Shell side/tube side performance
- Shell side/tube side construction
- Tubing, baffles, gaskets, etc. specs
- Sketches may be useful

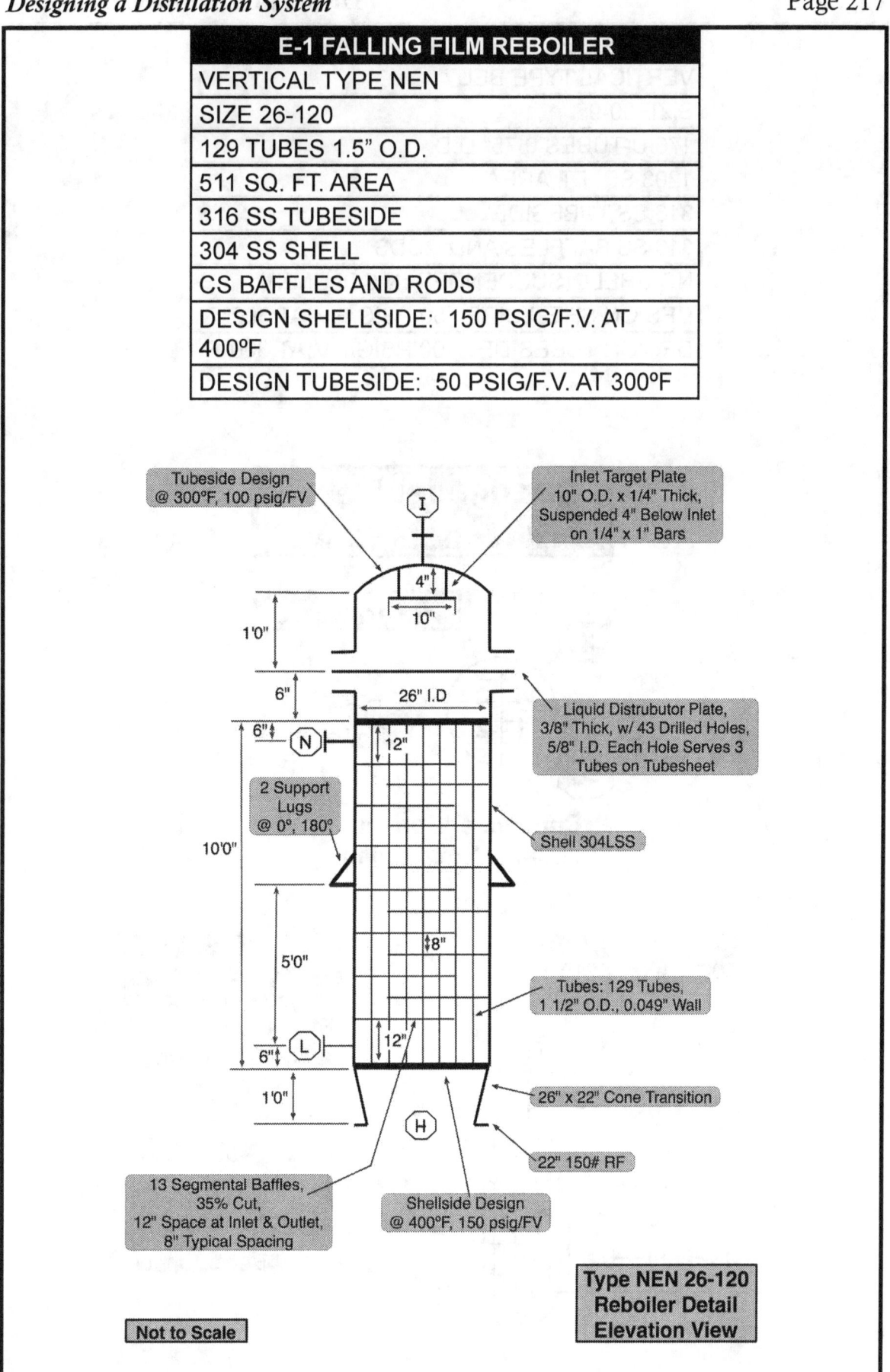

E-1 FALLING FILM REBOILER

VERTICAL TYPE NEN
SIZE 26-120
129 TUBES 1.5" O.D.
511 SQ. FT. AREA
316 SS TUBESIDE
304 SS SHELL
CS BAFFLES AND RODS
DESIGN SHELLSIDE: 150 PSIG/F.V. AT 400ºF
DESIGN TUBESIDE: 50 PSIG/F.V. AT 300ºF

Tubeside Design @ 300ºF, 100 psig/FV

Inlet Target Plate 10" O.D. x 1/4" Thick, Suspended 4" Below Inlet on 1/4" x 1" Bars

4"

10"

1'0"

6"

26" I.D

Liquid Distrubutor Plate, 3/8" Thick, w/ 43 Drilled Holes, 5/8" I.D. Each Hole Serves 3 Tubes on Tubesheet

6"

N

12"

2 Support Lugs @ 0º, 180º

Shell 304LSS

10'0"

8"

5'0"

Tubes: 129 Tubes, 1 1/2" O.D., 0.049" Wall

12"

L

6"

1'0"

26" x 22" Cone Transition

H

22" 150# RF

13 Segmental Baffles, 35% Cut, 12" Space at Inlet & Outlet, 8" Typical Spacing

Shellside Design @ 400ºF, 150 psig/FV

Type NEN 26-120 Reboiler Detail Elevation View

Not to Scale

E-2 VERTICAL KNOCKBACK CONDENSER
VERTICAL TYPE BEU
SIZE 30-96
176 U-TUBES 0.75" O.D.
1203 SQ. FT. AREA
316 SS TUBESIDE
316 SS BAFFLES AND RODS
NO SHELL (SUSPENDED IN COLUMN)
DESIGN SHELLSIDE: 50 PSIG/F.V. AT 300ºF
DESIGN TUBESIDE: 100 PSIG/F.V. AT 300ºF

Condenser Detail Elevation View

Not to Scale

K

| 4 | 3 |
| 1 | 2 |

Pass Number

J

Condenser Bonnet Divider & Nozzle Plan

Tubeside Design @ 300ºF 100 psig/FV

Type BEU 30-96 Condenser U-Bundle Inserted Through 30" Flange on Top of Distillation Column

2'6" I.D.

1'0"   J   K

45° Cone 2'6" I.D. 4'6" O.D.

1'0"   V   6"

1'0"

8'0"

7 Segmental Baffles, 40% Cut, @ 12" Spacing

P2

U-Bend Support

Vessel specification sheets

- Distillation columns
- Decanters
- Batch pots
- Product receivers
- Other tanks

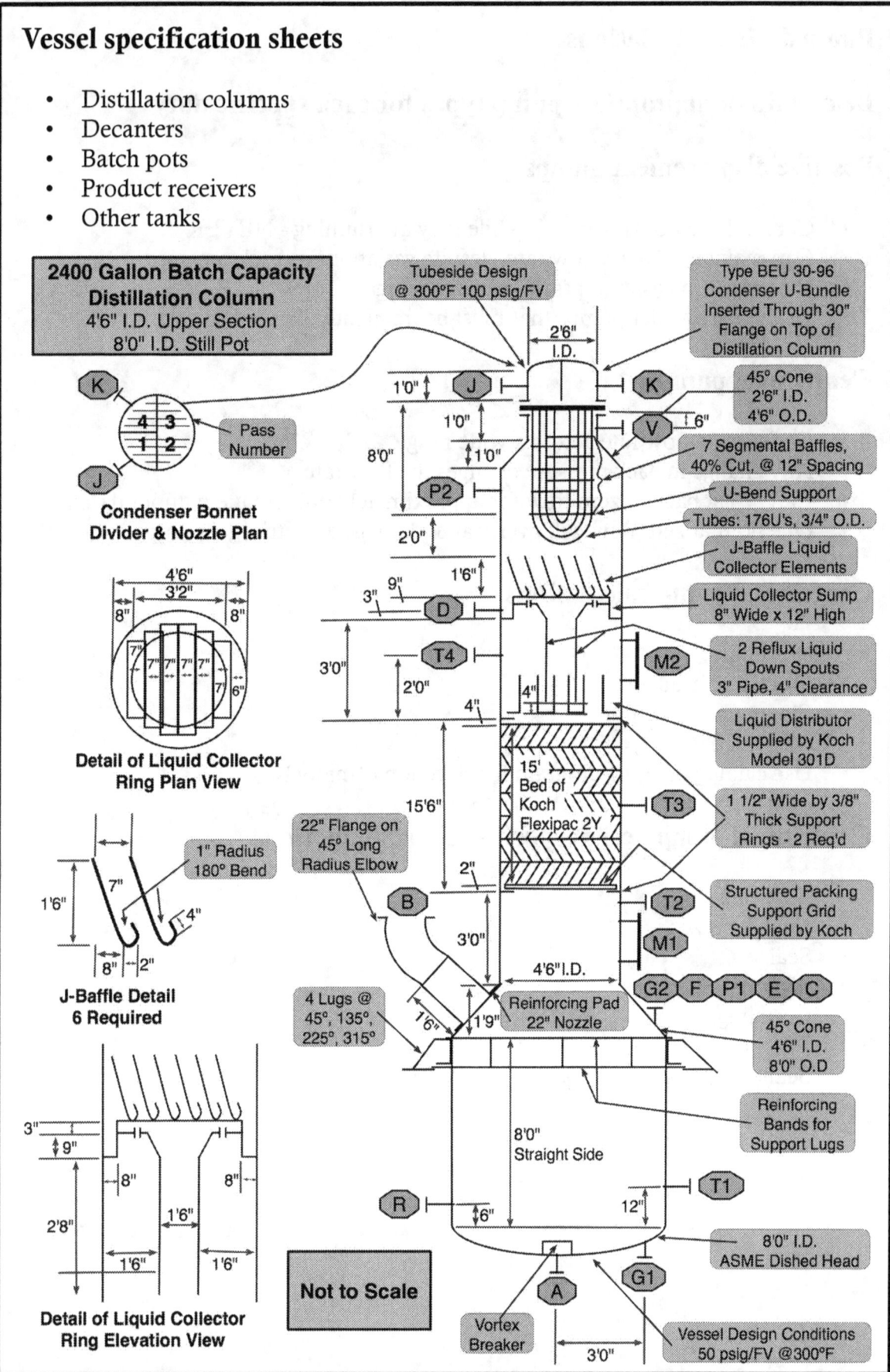

2400 Gallon Batch Capacity Distillation Column
4'6" I.D. Upper Section
8'0" I.D. Still Pot

Condenser Bonnet Divider & Nozzle Plan

Pass Number

Detail of Liquid Collector Ring Plan View

1" Radius 180° Bend

J-Baffle Detail 6 Required

Detail of Liquid Collector Ring Elevation View

Not to Scale

Tubeside Design @ 300°F 100 psig/FV

Type BEU 30-96 Condenser U-Bundle Inserted Through 30" Flange on Top of Distillation Column

45° Cone 2'6" I.D. 4'6" O.D.

7 Segmental Baffles, 40% Cut, @ 12" Spacing

U-Bend Support

Tubes: 176U's, 3/4" O.D.

J-Baffle Liquid Collector Elements

Liquid Collector Sump 8" Wide x 12" High

2 Reflux Liquid Down Spouts 3" Pipe, 4" Clearance

Liquid Distributor Supplied by Koch Model 301D

1 1/2" Wide by 3/8" Thick Support Rings - 2 Req'd

15' Bed of Koch Flexipac 2Y

Structured Packing Support Grid Supplied by Koch

22" Flange on 45° Long Radius Elbow

4 Lugs @ 45°, 135°, 225°, 315°

Reinforcing Pad 22" Nozzle

45° Cone 4'6" I.D. 8'0" O.D

Reinforcing Bands for Support Lugs

8'0" Straight Side

8'0" I.D. ASME Dished Head

Vortex Breaker

Vessel Design Conditions 50 psig/FV @300°F

Pump design calculations

Be certain of appropriate pump types for each specific duty

Positive displacement pumps:

- Cannot have flow restricted, while they are running - NEVER
- Can maintain a fixed flow rate, despite variations in back pressure
- Can be much easier to prime than centrifugal pumps
- Useful for transfer pumps that must be frequently drained

Centrifugal pumps:

- Can vary in flow rate across a wide range
- Any change in back pressure changes the flow rate
- Can be cut back to zero flow for approximately one minute without damage
- One hour at zero flow will overheat & damage a centrifugal pump

Centrifugal pump calculations

First determine:
- Discharge head
- Flow rate
- NPSHA
- Use catalog or equal program for selection of models

Centrifugal pump design specification completion

- Materials
- Impellor size
- Seals
- Motors
- Couplings
- Baseplates
- Sealless options

Positive displacement pump calculations

First determine:
- Discharge head (maximum)
- Flow rate
- NPSHA
- Use catalog or equal program for selection of models

Positive displacement pump design specification completion

- Materials
- Seals
- Motors
- Speed reducers
- Variable speed drives
- Couplings
- Baseplates

Equipment Data – For Display On P&Id	
P-1 Feed Pump	**P-3 Hot Oil Circulation Pump**
Oberdorfer Model 13500	Dean Brothers Model RMA 5000
Bronze Construction Gear Pump	Size 3 X 4 – 8.5 Centrifugal Pump
Teflon Packing	Steel Construction For 600ºF Service
42 GPM At 20 PSI Head	Fabricated Steel Baseplate
2 HP 1750 RPM 460 V Motor	450 GPM At 50 Ft Head
1800 X 900 RPM Gear Reducer	Fluid Density 0.84 G./Cc
P-2 Reboiler Circulation Pump	NPSHR = 6 Ft
Goulds Model 3196 MTX	7.30" Impellor
Size 3 X 4 – 8 – A70 Centrifugal Pump	10 HP 1750 RPM Motor
316 Ss Construction	**P-4 Terpenes Pump**
Cast Iron Baseplate	Oberdorfer Model 13500
Dura X-200 Cartridge Double Seal	Bronze Construction Gear Pump
7353 Flush Plan	Teflon Packing
450 GPM At 55 Ft Head	42 GPM At 20 PSI Head
Fluid Density 0.86 G./Cc	2 HP 1750 RPM 460 V Motor
NPSHR = 6 Ft	1800 X 900 RPM Gear Reducer
8.00" Impellor	
10 HP 1750 RPM 460 V Motor	

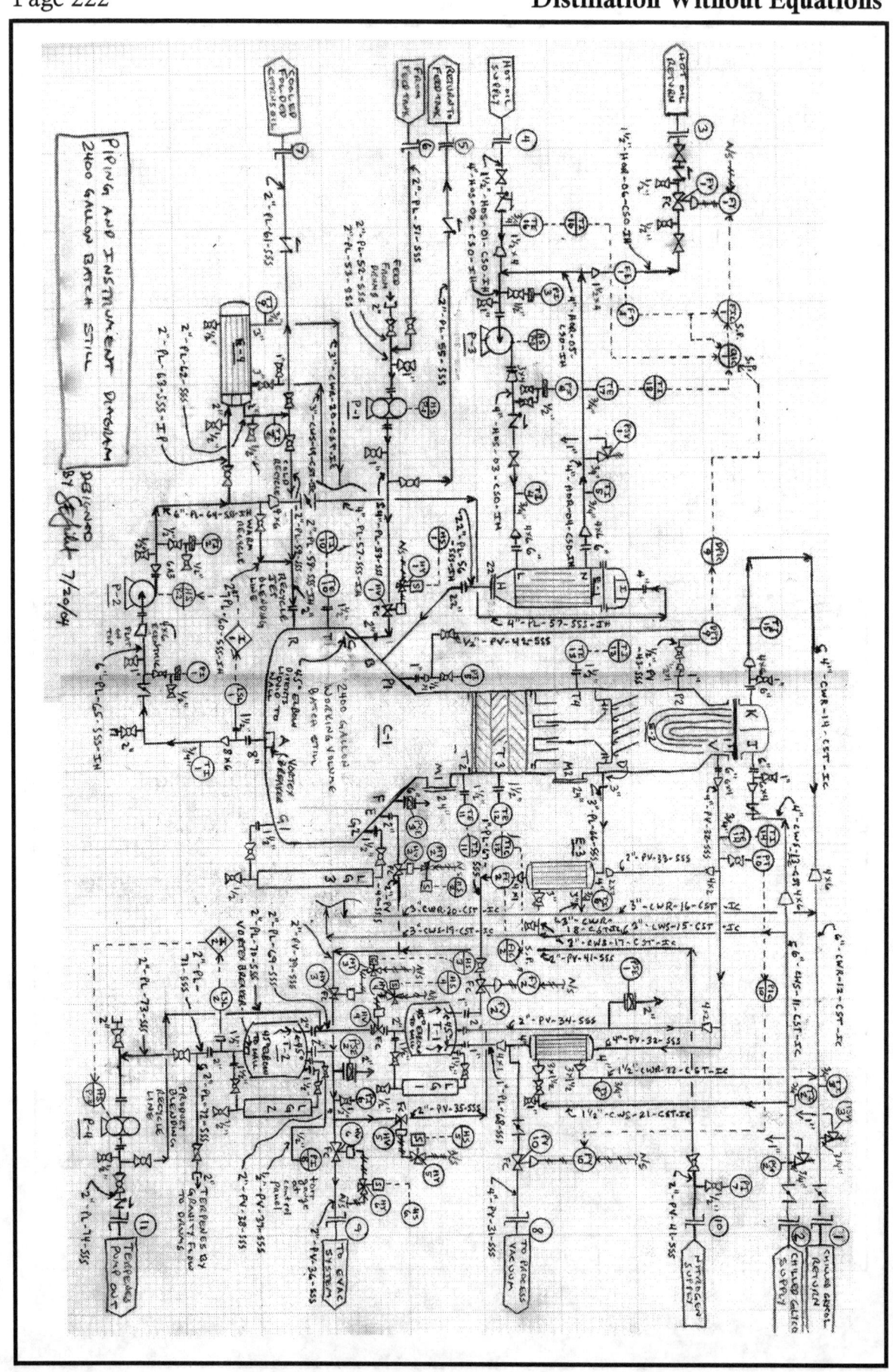

Create the new distillation system P&ID

- Draw by hand, or with a computer drafting or graphics design program
- Duties & connection sizes must be known
- What valves, instrumentation are required to operate this plant safely?
- Consider how to fill the system, start it, shut it down, & empty it

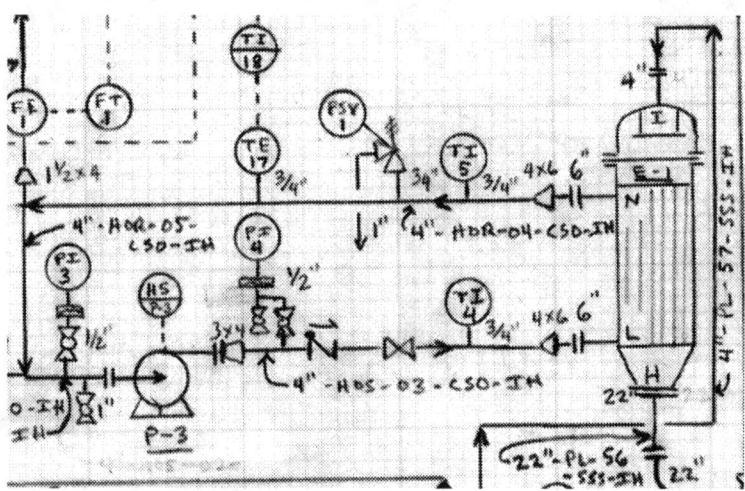

How to create a new P&ID

- Start with equipment & main process pipe lines from the block flow diagram
- Add more detailed process piping
- Add utility supply & return piping
- Add instruments to measure & control the process & utilities
- Add valves for control, protection, or isolation
- Add pressure safety equipment for any section containing a vessel that can be blocked in
- Create & design control loops & interlocks

Start with a logical arrangement of equipment from the block flow diagram

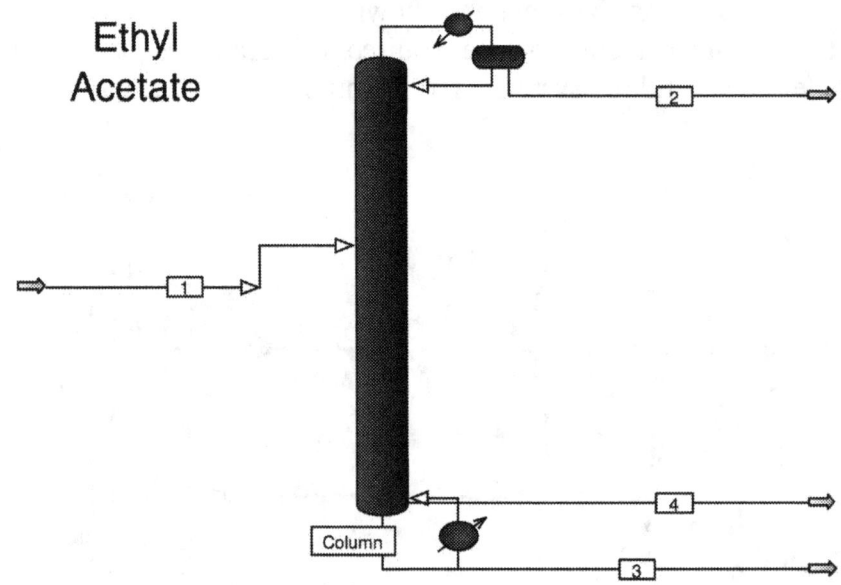

Add main process pipe lines from the block flow diagram

Add more detailed process path piping

Add utility supply & return & vent piping

Add measurement instruments for flow, pressure, temperature, level

Review the pipe line map to make certain that the equipment can be filled, emptied, & recycled

Add valves for control, protection, & isolation

Add safety equipment:

- Pressure relief PSV or PSE for any section containing a vessel that can be blocked in
- High level switches & low level switches
- High pressure switches if appropriate

Control strategy must be determined for:

- Feed input
- Heat input
- Column product purity control
- Column bottom effluent
- Reflux
- Column top distillate
- Other feed, product, or internal streams

Control loops must be created & designed to carry out selected control strategy

Add process indicators & safety interlocks

- Indicators added for information or data recording purposes
- Interlock devices included for protection of pumps & overall system safety

Continue development of P&ID after creation of all pipe lines, instruments, & controls:

- Instruments tagged & numbered for lists & specifications
- Every section of pipe tagged with size - fluid description - material code - insulation

Conduct P&ID operability review

Make certain that the system allows for:
- Filling
- Starting up
- Sampling
- Shutting down
- Emptying

Conduct P&ID environmental review:

- No uncontrolled vent emissions
- Seals on pumps & instruments up to good practices
- Means included to stop accidental spills or vent releases
- Plugs or blind flanges to prevent accidental releases
- Containment requirements for spills in the area

Conduct preliminary P&ID safety review

- Use HAZOP procedure or other safety review procedure
- Identify possible abnormal performance
- Include mitigating measures:
    - check valves, interlocks, level switches, pressure safety valves, burst disks, extra valves or caps on drains & vent connections

Revise P&ID in accordance with recommendations of reviews:

- Operability
- Environmental
- Safety

Back-check the revised P&ID to make certain that:

- Review comments have been incorporated correctly
- No new problems have been created

Complete the P&ID development process

By attaching summary information to the P&ID, about equipment on the P&ID:
- Pumps
- Distillation columns
- Decanters
- Vessels
- Heat exchangers

EQUIPMENT DATA – FOR DISPLAY ON P&ID	
E-1 FALLING FILM REBOILER	**E-3 DISTILLATE COOLER**
VERTICAL TYPE NEN	STANDARD SHELL & TUBE
SIZE 26-120	YOUNG RADIATOR CORP.
129 TUBES 1.5" O.D.	MODEL SSF – 810 – ER – 1P
511 SQ. FT. AREA	3/8" O.D. TUBES 91" LONG
316 SS TUBESIDE	316 SS TUBESIDE
304 SS SHELL	CS SHELLSIDE
CS BAFFLES AND RODS	DESIGN 100 PSIG AT 300ºF
DESIGN SHELLSIDE: 150 PSIG/F.V. AT 400ºF	**E-4 FOLDED CITRUS OIL COOLER**
DESIGN TUBESIDE: 50 PSIG/F.V. AT 300ºF	STANDARD SHELL & TUBE
E-2 VERTICAL KNOCKBACK CONDENSER	YOUNG RADIATOR CORP.
VERTICAL TYPE BEU	MODEL SSF – 810 – ER – 4P
SIZE 30-96	3/8" O.D. TUBES 91" LONG
176 U-TUBES 0.75" O.D.	316 SS TUBESIDE
1203 SQ. FT. AREA	CS SHELLSIDE
316 SS TUBESIDE	DESIGN 100 PSIG AT 300ºF
316 SS BAFFLES AND RODS	**E-5 VENT CONDENSER**
NO SHELL (SUSPENDED IN COLUMN)	STANDARD SHELL & TUBE
DESIGN SHELLSIDE: 50 PSIG/F.V. AT 300ºF	YOUNG RADIATOR CORP.
DESIGN TUBESIDE: 100 PSIG/F.V. AT 300ºF	MODEL SSF – 806 – ER – 1P
	3/8" O.D. TUBES 55" LONG
	316 SS TUBESIDE
	CS SHELLSIDE
	DESIGN 100 PSIG AT 300ºF

Develop instrument list

- Instrument list will be extracted from the P&ID
- Can be created automatically or by hand
- Automation can be better if working from scratch, but needs non-interfering humans to accept its automatically generated numbering

Instrument list contents (for each instrument)

- Quantity
- Tag #
- Type
- Service
- Range
- Connections
- Material

Instrument list categories

- Part 1: Field instruments, electric & mechanical
- Part 2: Control panel devices, electrical & pneumatic
- Part 3: Control valves & actuated valves

		Part 1: Field Instruments, Electric & Mechanical				
Qty	Tag #	Type	Service	Range	Connections	Material
	DPT 9	2 Port DP	Column Vapor	0-25 mmHg	1/2" NPT	316
	FE 1	Orifice DP	Hot Oil	0-40 GPM	1 1/2" 150#	CS / SS trim
	FE 2	Coriolis Mass Flow w/Density E&H 63F	Terpene Distillate	0-9000 lb/hr	1" 150#	316 / 904L
	LSL 1	E&H M FTL 51	Citrus Oil	Fail Empty	1 1/2" 150#	316
	LSL 2	E&H M FTL 51	Terpenes	Fail Empty	1 1/2" 150#	316
	PI 1	PI / Dia. Seal	P-2 Suction	30"Vac-0-15	1/2" NPT	316
	PI 2	PI / Dia. Seal	P-2 Discharge	30"Vac-0-60	1/2" NPT	316
	PI 3	PI / Dia. Seal	P-1 Suction	0-100 psig	1/2" NPT	Brass
	PI 4	PI / Dia. Seal	P-1 Discharge	0-100 psig	1/2" NPT	Brass
	PI 5	Pressure Guage	Column Bot	30"Vac-0-15	1/2" NPT	316
	PI 7	Pressure Guage	Nitrogen	0-30 psig	1/2" NPT	316
	PI 8	Torr Pnl Guage	T-2 Abs Press	0-30 torr abs	1/2" NPT	316
	PSE 1	Burst Disk	Full Vac	Set 50 psig	2" 150#	316
	PSE 2	Burst Disk	Full Vac	Set 50 psig	2" 150#	316
	PSE 3	Burst Disk	Full Vac	Set 50 psig	6" 150#	316
	PSV 1	Liq Expansion	Hot Oil	Set 150 psig	3/4"x1" NPT	CS / SS trim
	PSV 2	Liq Expansion	Chilled Glycol	Set 100 psig	3/4"x1" NPT	CS / SS trim
	PSV 3	Liq Expansion	Chilled Glycol	Set 100 psig	3/4"x1" NPT	CS / SS trim
	PT 10	Absolute Press	Cond Vent	0-30 torr abs	1/2" NPT	316
	TE 10	RTD 100 ohm	Still Pot Temp	0-300 F	1 1/2" 150#	316
	TE 11	RTD 100 ohm	Col Bot Temp	0-300 F	1 1/2" 150#	316
	TE 12	RTD 100 ohm	Col Mid Temp	0-300 F	1 1/2" 150#	316
	TE 13	RTD 100 ohm	Col Top Temp	0-300 F	1 1/2" 150#	316
	TE 14	RTD 100 ohm	Col Vent T	0-300 F	3/4" NPT	316
	TE 16	RTD 100 ohm	Hot Oil Supply	50-450 F	3/4" NPT	Brass
	TE 17	RTD 100 ohm	Hot Oil Return	50-450 F	3/4" NPT	Brass
	TI 1	Dial Thermometer	Liquid Out of C-1 to P-2		3/4" NPT	316
	TI 15	Dial Thermometer	Chilled Glycol		3/4" NPT	Brass
	TI 2	Dial Thermometer	Chilled Glycol		3/4" NPT	Brass
	TI 3	Dial Thermometer	Chilled Glycol		3/4" NPT	Brass
	TI 4	Dial Thermometer	Hot Oil		3/4" NPT	Brass
	TI 5	Dial Thermometer	Hot Oil		3/4" NPT	Brass
	TI 6	Dial Thermometer	Chilled Glycol		3/4" NPT	Brass
	TI 7	Dial Thermometer	Cooled Folded Citrus Oil		3/4" NPT	316
	TI 8	Dial Thermometer	Chilled Glycol		3/4" NPT	Brass
	TI 9	Dial Thermometer	Chilled Glycol		3/4" NPT	Brass

Qty	Tag #'s	Type	Service	Range	Connections	Material
		Part 2: Control Panel Devices, Electrical & Pneumatic				
	DPIC 9	DP Controller	Column DP	0-25 mmHG		
	FIC 1	Flow Controller	Hot Oil Flow	0-40 GPM		
	FIC 2	Flow Controller	Terpene Distillate Flow	0-9000 lb/hr		
	HS 1	Switch 115V	Citrus Oil Feed	Open-Close		
	HS 2	Switch 115V	Nitrogen to C1	Open-Close		
	HS 3	Switch 115V	Nitrogen to T2	Open-Close		
	HS 4	Switch 115V	T1 Liq to T2	Open-Close		
	HS 5	Switch 115V	T2 Vent to T1	Open-Close		
	HS 6	Switch 115V	T2 to Evac Sys	Open-Close		
	HS P-1	Motor Switch	Feed Pump	Run-Stop		
	HS P-2	Motor Switch	Reboiler Circ	Run-Stop		
	HS P-3	Motor Switch	Hot Oil Circ	Run-Stop		
	HS P-4	Motor Switch	Terpenes Out	Run-Stop		
	I 1	P2 Interlock	LSL1 Stops P2			
	I 2	P4 Interlock	LSL2 Stops P4			
	PIC 10	Pres Controller	Condenser Vent Vacuum	0-30 torr abs		
	QIC 1	Heat Controller	Btu Input to Reboiler	0-1M Btu/hr		
	TI 10	T Indicator	Still Pot	0-300 F		
	TI 11	T Indicator	Column Bot	0-300 F		
	TI 13	T Indicator	Column Top	0-300 F		
	TI 14	T Indicator	Column Vent	0-300 F		
	TI 16	T Indicator	Hot Oil Supply	50-450 F		
	TI 17	T Indicator	Hot Oil Return	50-450 F		
	TIC 12	T Controller	Column Mid	0-300 F		

			Part 3: Control Valves & Actuated Valves			
Qty	Tag #	Type	Service	Range	Connections	Material
	FV 1	Globe Control	Hot Oil 130-450 F	0-40 GPM	1 1/2" 150#	CS/SS Trim
	FV 2	V-Ball Control	Terpene Grav. Flow	0-22 GPM	1" 150#	316
	FY 1	I/P positioner	3-15 psig output	4-20 ma	NPT	Brass
	FY 2	I/P positioner	3-15 psig output	4-20 ma	NPT	Brass
	HV 1	Ball Vlv On-Off	Citrus Oil Feed	0-15 psig	2" SW	316
	HV 2	Ball Vlv On-Off	Nitrogen	0-15 psig	2" SW	316
	HV 3	Ball Vlv On-Off	Nitrogen	0-15 psig	2" SW	316
	HV 4	Ball Vlv On-Off	T-1 Liquid to T-2	0-15 psig	2" SW	316
	HV 5	Ball Vlv On-Off	T-2 Vent to T-1	0-15 psig	2" SW	316
	HV 6	Ball Vlv On-Off	T-2 to Evac System	0-15 psig	2" SW	316
	HY 1	Solenoid 3-port	Energize to Open	115 VAC	NPT	Brass
	HY 2	Solenoid 3-port	Energize to Open	115 VAC	NPT	Brass
	HY 3	Solenoid 3-port	Energize to Open	115 VAC	NPT	Brass
	HY 4	Solenoid 3-port	Energize to Open	115 VAC	NPT	Brass
	HY 5	Solenoid 3-port	Energize to Open	115 VAC	NPT	Brass
	HY 6	Solenoid 3-port	Energize to Open	115 VAC	NPT	Brass
	PV 10	V-Ball Control	Process Vacuum		4" 150#	316
	PY 10	I/P positioner	3-15 psig output	4-20 ma	NPT	Brass

Pipe line list

- Extracted from P&ID
- All lines listed by type
- Process vapor
- Process liquid
- Steam
- Nitrogen
- Cooling water

Create a workable system layout

Create plan, elevation, & 3D layout views:
- Where will everything sit?
- How much space will it take?
- How will people access the area?
- How will equipment be accessed?
- How will the system be assembled?
- How will the system be maintained?

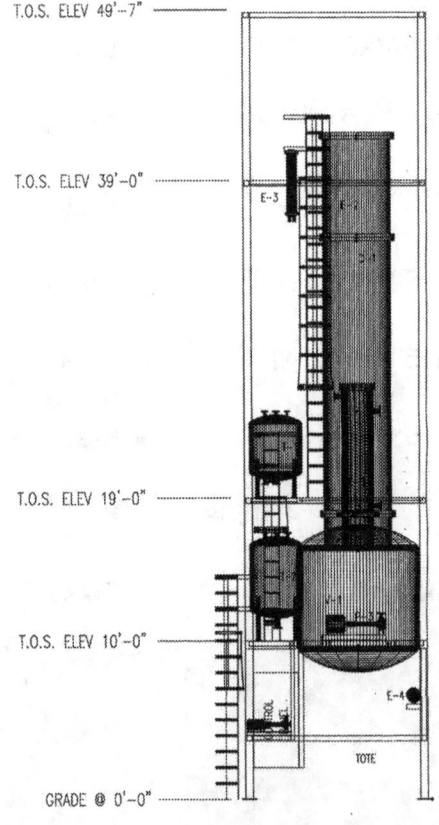

ELEVATION VIEW LOOKING WEST

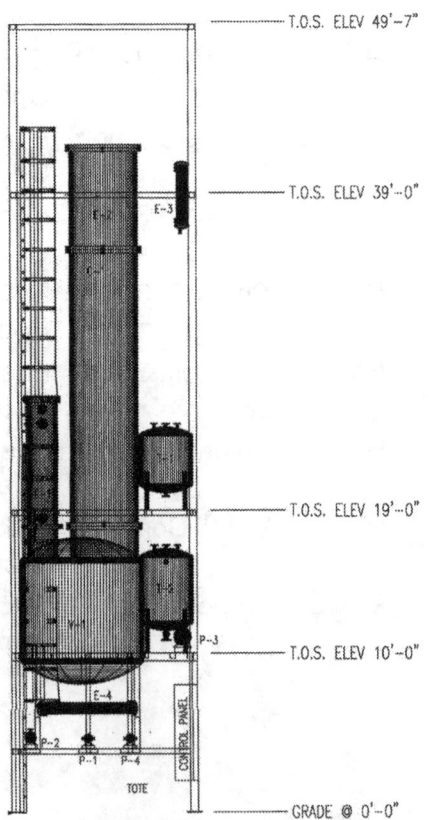

ELEVATION VIEW LOOKING NORTH

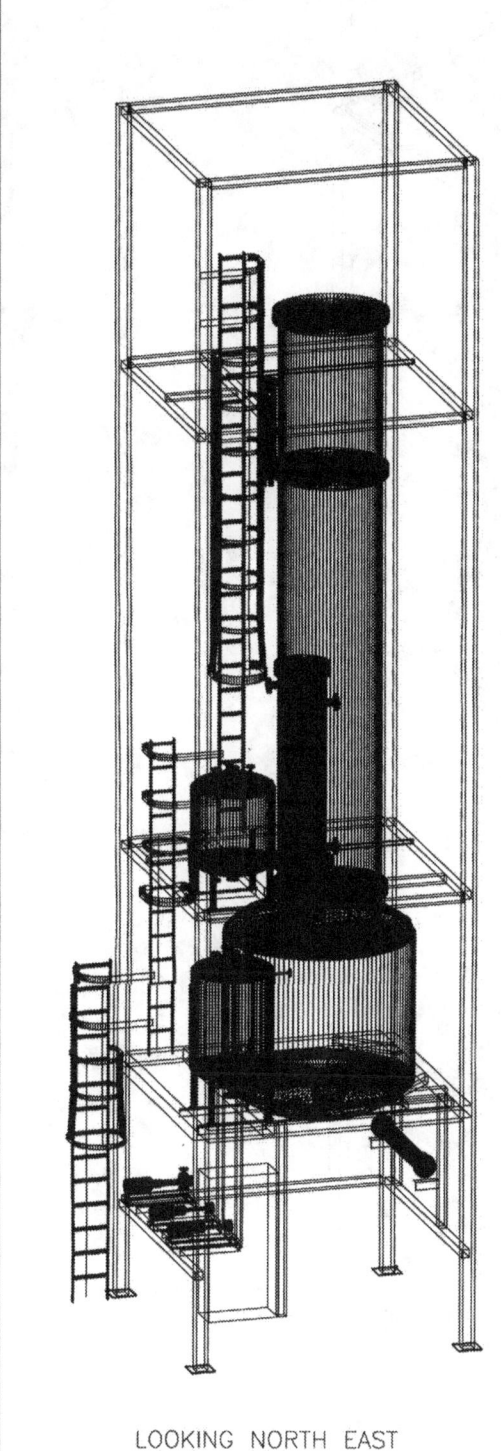

LOOKING NORTH EAST

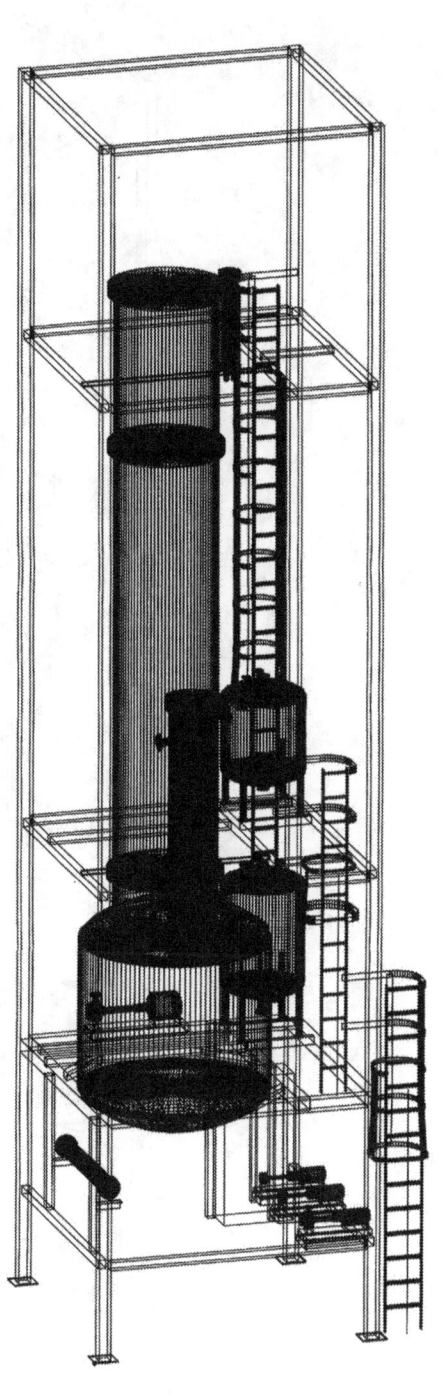

LOOKING SOUTH WEST

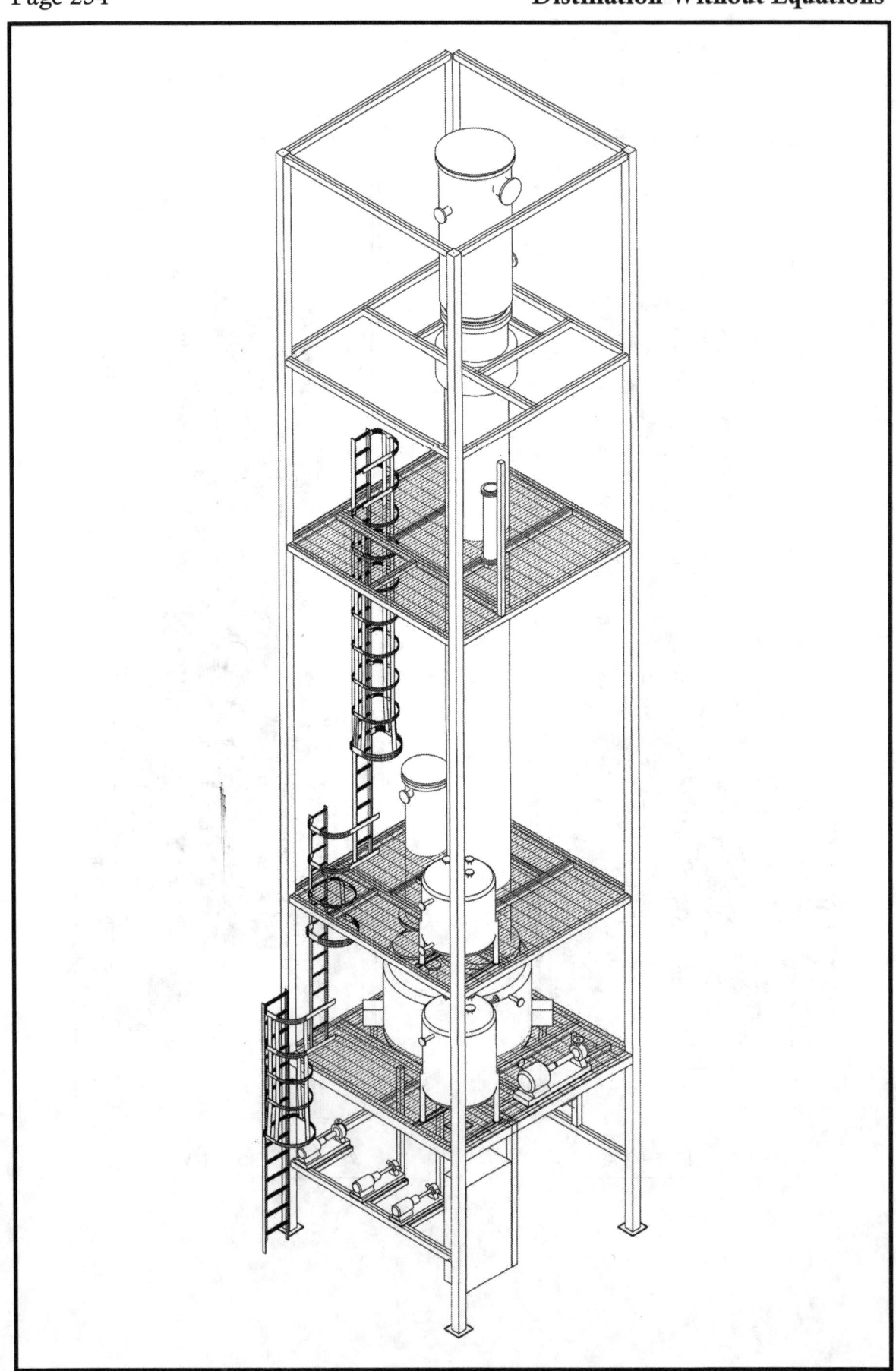

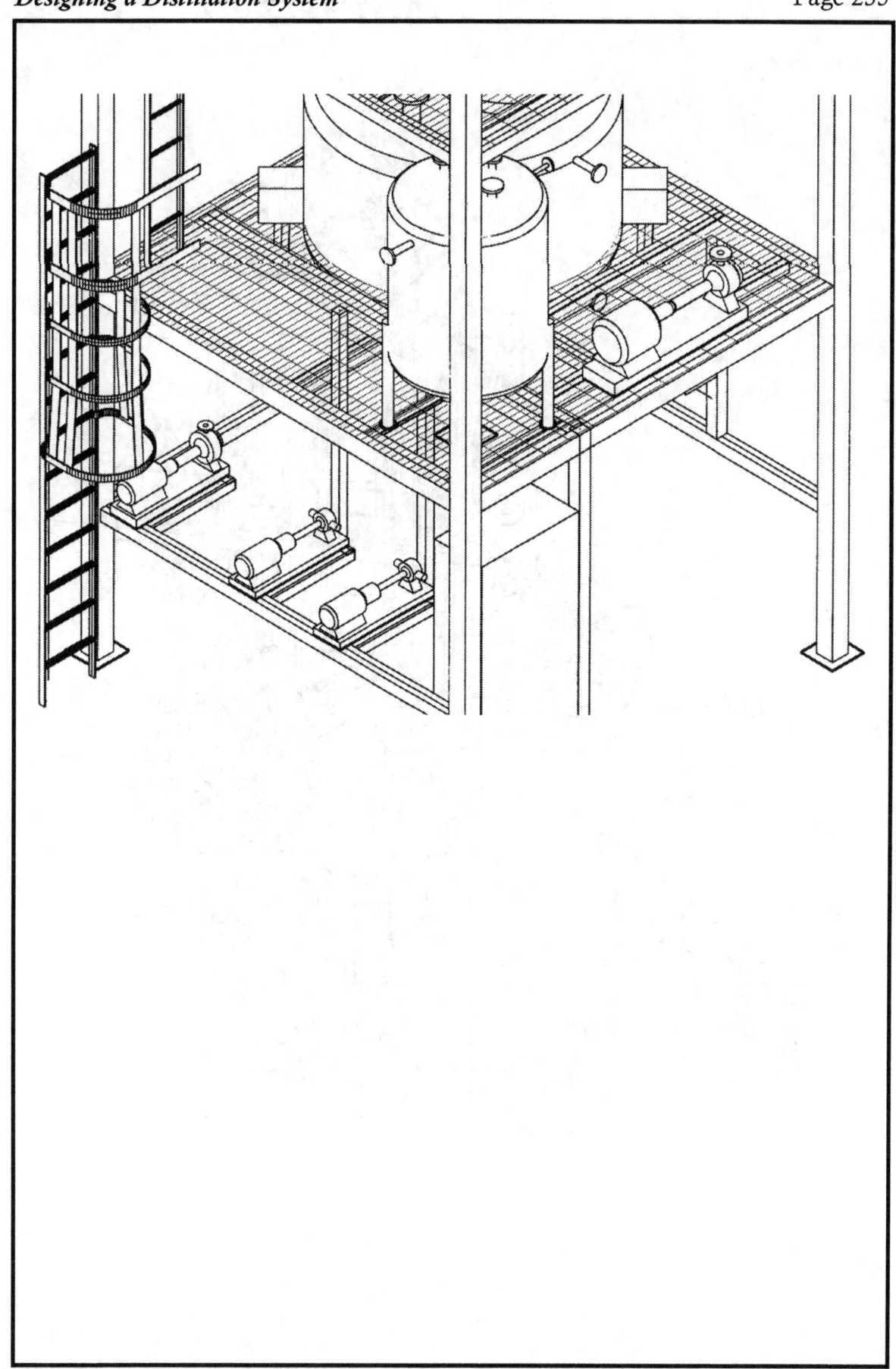

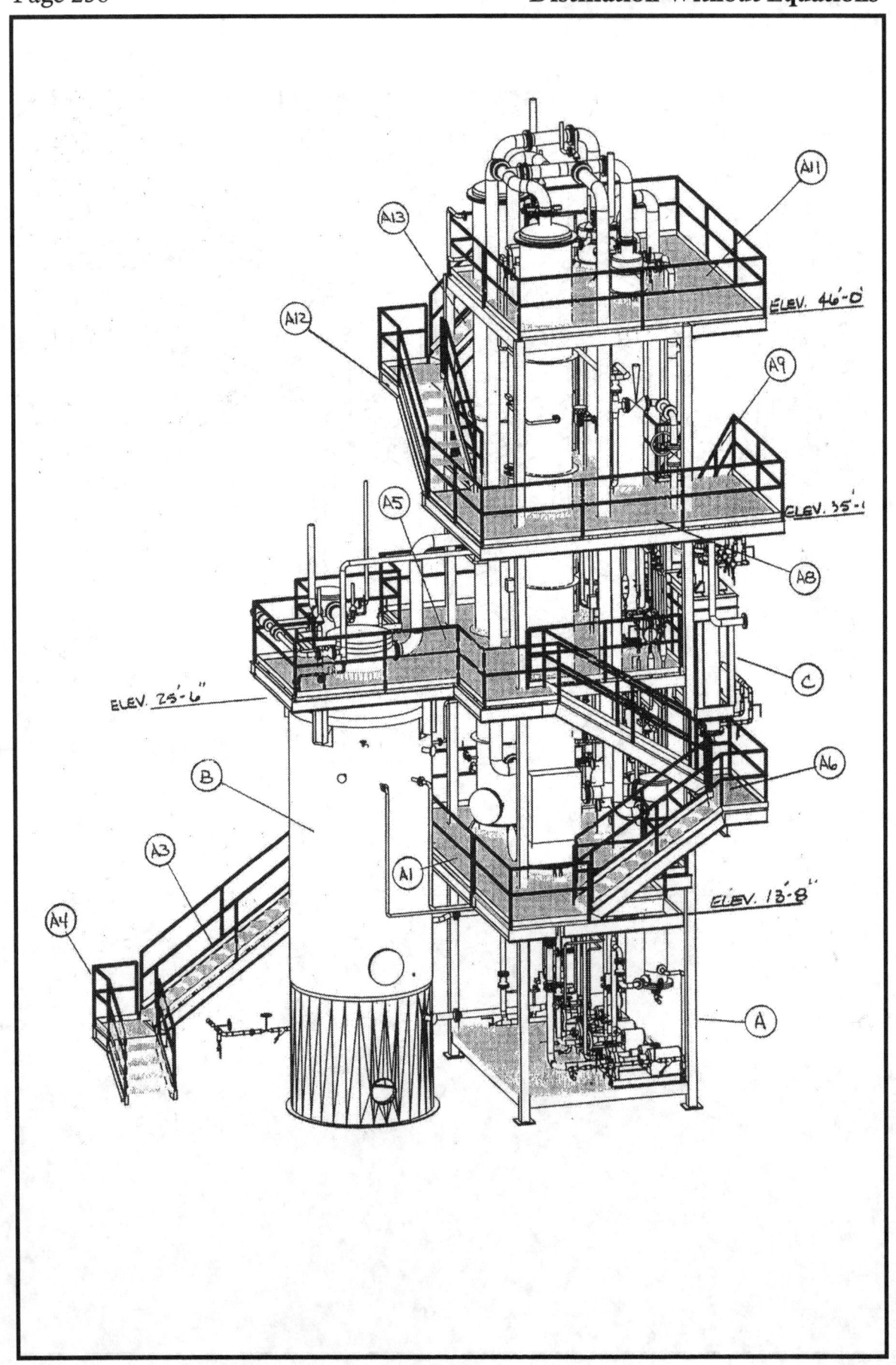

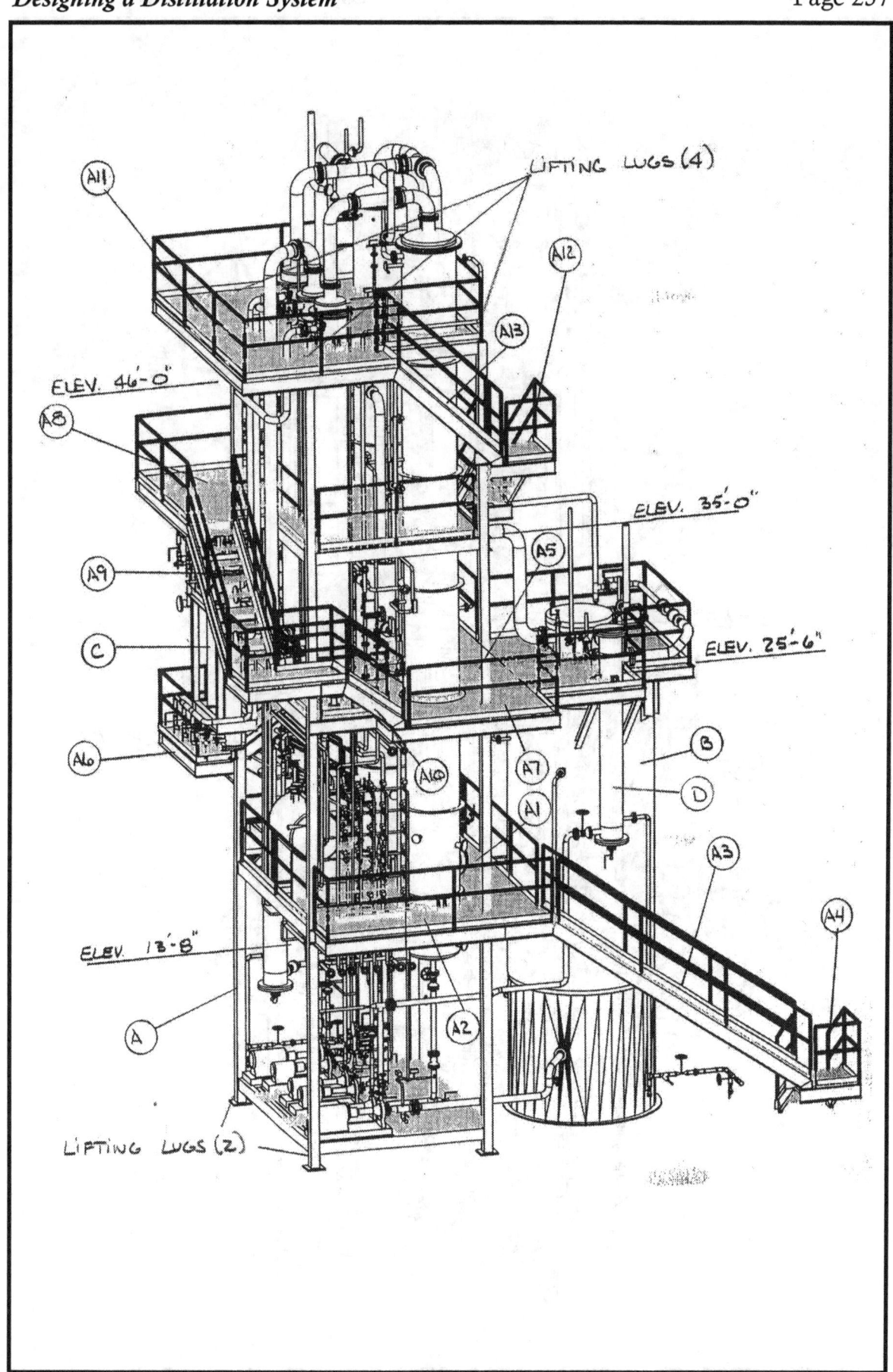

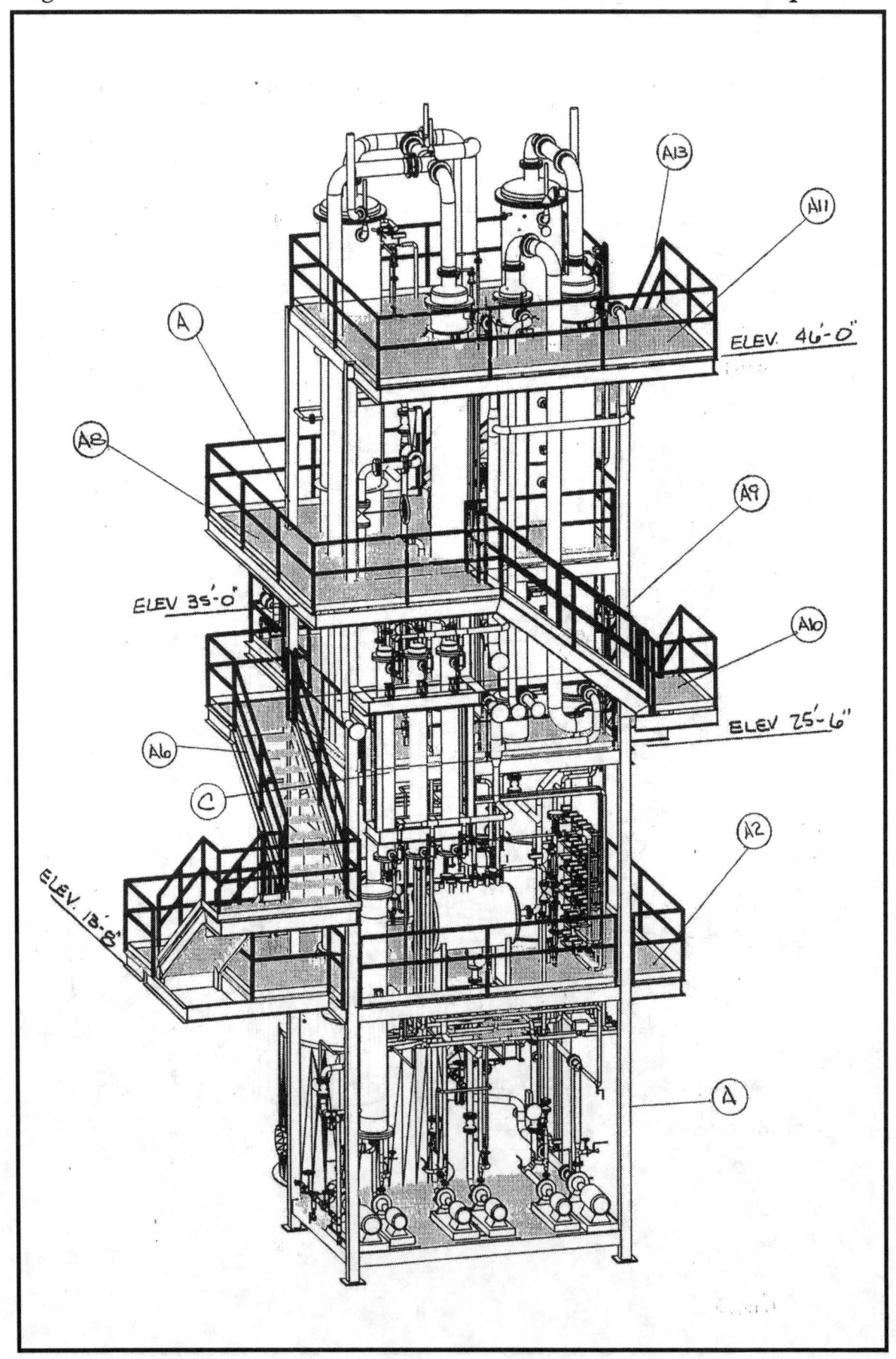

ELEV. 46'-0"

ELEV 35'-0"

ELEV 25'-6"

ELEV. 13'-8"

Big Bang Ends

Proceed to detail engineering, fabrication, & assembly

- Detail engineering includes everything!
- Control panel, instruments, vessels, gaskets, etc.
- Follow P&ID to the letter - it is your Bible!
- Mark up redline versions to correct any mistakes or record changes to the P&ID
- Issue updated versions for construction
- Field verify as-built version of P&ID

Pay attention to how it will look! Distillation systems will be seen!

Your Author and Instructor:

Steve Licht, BS Chemical Engineering (1979), University of Illinois at Chicago, began his chemical industry career in 1973 as a pilot plant operator and technician making laboratory vapor liquid equilibrium measurements. He has presented live seminars from the **"Distillation Without Equations"** seminar series to groups of engineers, scientists, plant operators, and supervisors, in Ireland, Puerto Rico, and India, in a technology transfer effort accompanying the migration of chemical and pharmaceutical manufacturing facilities to these regions. Steve has designed over 80 distillation systems placed into operation in diverse industries (with a concentration in bulk pharmaceuticals) in North and South America, Europe, and Asia. He now provides engineering consulting services for distillation systems and control systems, including testing and validation of modern computer process control systems, and presents live seminars from the **"Distillation Without Equations"** series as demand warrants. To request further information or to arrange a live seminar, email your request to distillbooks @yahoo.com.

www.ingramcontent.com/pod-product-compliance
Lightning Source LLC
Chambersburg PA
CBHW081112170526
45165CB00008B/2420